Moral Tales
and Meditations

Moral Tales and Meditations

Technological Parables and Refractions

Michael Joyce

Afterword by Hélène Cixous

State University of New York Press

Published by
State University of New York Press, Albany

For information, address State University of New York Press,
90 State Street, Suite 700, Albany, NY 12207

Production by Dana Foote
Marketing by Patrick Durocher

Library of Congress Cataloging-in-Publication Data

Joyce, Michael, 1945–
Moral tales and meditations : technological parables and refractions
/ Michael Joyce ; with an afterword by Hélène Cixous.
p. cm.
Includes bibliographical references.
ISBN 0-7914-5155-0 (alk. paper)
ISBN 0-7914-5156-9 (pb: alk. paper)
1. Technology and civilization—Moral and ethical aspects—Fiction.
2. Technology and civilization—Moral and ethical aspects.   3. Mass media—
Moral and ethical aspects—Fiction.   4. Mass media—Moral and ethical aspects.
5. Didactic fiction, American.   6. Parables.   I. Title.

PS3560.O885 M6 2001
813'.54—dc21
2001020949

10 9 8 7 6 5 4 3 2 1

In memory of Alicia McCall (1927–1999) these stories' first reader, and who appears throughout in various guises

Cumpleaños

from the balcón the fuegos artificiales
No, from the balcony the salt scent
of the sea, the murmur turned to whisper
at this distance, and lovers promenade
along la costa, along the foothills shadows
of dusk, all the voices mixed by breezes
and the years where? Here, of course, before
your eyes, the slim fellow with eyes como
un fox (the word is zorro, the kids are watching
the television and outside in the darkness
beyond the tree line and highway sea nettles
swirl in their silk dresses in the dark cove).

From the balcony the years flare
like fireworks in the violet basin
of the valley and the sea is no different,
another violet curve, outstretched
palm in the dying light of the mercado,
and on the sea candles bobbing, no the
phosphorescent sheen of the surf,
moonlight, your fine cheekbones
and crimson lips, a matador would
fall for someone like you, remove
the silk from you or him or both,
the two of you built a house (you
always built a house) that was the
talk of the county, the sun captured
there in the plateglass splashing
onto the polished hood of the
dark sedan he drove home today
so he could see you perched there
in the reflected light, over all these
years, the fire still there, the promise

CONTENTS

Contents

THREE LAST PIECES

INTRODUCTION

I have never considered myself a short story writer (and it may well be that some readers will think that these short-short stories confirm that opinion) but the stories here came to me almost as visions and in a way, over the course of a spring and summer, I could not resist. From the beginning I understood them to be moral tales, at least in some part parables of technology. In fact they arrived almost formulaically, like sonnets, that is, each with a requisite mention of technology, each with one or another character trying to hold on to (or regain) her humanity, each with a haiku quality, a lyric turn which deflects its trajectory, or so one hopes, just before the point of sentimentality.

This is not to say that I set out to write them formulaically but rather that they offered up these elements in a way that at least partially explained why, after not working in this form through thirty some years of a writing life, I suddenly could not so much finish one story before another presented itself: here a feminist parachutist on the run, there the toothbrush arriving via Federal Express.

A winter and another spring passed and a few of the stories were published, or about to be, when in a second winter the notion came upon me of collecting these short tales with a parallel set of equally short meditations, not commentaries on the stories exactly (or at all) as much as mirrors to them.

I have as a model (one I am certainly unworthy of) Italo Calvino's profound *Six Memos for the Next Millennium,* although

what they model in this instance is not the scope—since my meditations here are nearly (or merely) memoirs without either the intellec-
tual range or historical grounding of Calvino's lectures—but rather
the inclination, a writer's memoir at the edge of some future. As such,
the meditations here continue an effort to construct what I have
called elsewhere "theoretical narratives." (If this were a more rigorous
text I could perhaps pause to note how rich the theoretical landscape
populated by narrative figures, from Deleuze and Guattari, to my
beloved Cixous, from Geertz to Haraway. Indeed the narrative of theory is both a genre and itself a meta-theoretical insight.)

Calvino might also seem a model for the short fictions themselves, which I dare to hope have at least some of the grace and a jot
of the wit of his work. Even without their mirror pieces, these stories
are curiously hypertextual, at least to the extent that they skip their
own bounds within a short span, spilling over into each other and
events around them in a way electronic hypertext fictions do as well.
On the other hand I hope the stories also summon the (arti-)factual
world a little like their predecessor texts did, not only Calvino but the
short stories of the great club of mordant and mischievous (but my
oh my oh-so-male) writers, especially Barth, Barthelme, Coover, and
Gass—who were my distant mentors when I was a young writer and
postmodern fiction meant something other than what many of us
have come to write in recent years.

One might well ask why a hypertext writer comes to write a series of short and linear (print) fictions. One answer is that they become not so much a respite from work with electronic forms as a con-
tinuity with them. (One of these stories in fact emerged in segments
from the memory of a character in my web fiction, *Twelve Blue.*) Why
this happened now, however, I wish I knew exactly.

Partly I think I have been interested in how we learn the world
by mouth, trying out new terms, virtual reality or user-friendly for
instance, in much the same way that an infant surprises herself by

whistling at will by forming her lips just so and breathing; or, as one of the meditations recounts, how a child boldly uses a new word, television for instance, well before seeing one. Technology, I have suggested in these and other fictions and meditations, is a lens to mortality. We see ourselves emerging and disappearing simultaneously.

Perhaps that will have to do for why these pieces appeared. "Things come and go" is itself a parable. There was a spring and these stories came, and two winters later I polished stones into uncertain mirrors. And we all lived, happily, ever after.

Afterward: Of course nothing ends that way, not even, it turns out, this collection. For good or for ill (but chiefly to add a little more "spine," both theoretical and in the sense the book trade understands that word), the publishers sought a bit more overtly theoretical and hypertextual heft here at the end, a firmer indication for the fainthearted (or feint-hardened) reader of what direction has been taken. Or, better still perhaps, what direction she or he has been taken in. Or again, worse, to assure him or her that she or he has not been taken in.

In any case there follow three last pieces here. I could have said new, I suspect, but I am anymore wary of newness in anyone, myself especially. I likewise considered entitling the section "three more pieces," thus suggesting the figure of a puzzle. While the "pieces" part of that phrase correctly reflects the current state of my thinking, it is the "more" in it that would have misled, at least to the extent that it suggests a faith in progress, an eventual hope that we will "solve" the puzzle of this emerging cyberculture or the lives we lead in it and which have led to it. The truth is that in the scheme of things, and in my life as it is now, these last pieces do seem the end of something. So last they are and will be, if not lasting.

In the same wise and for much the same reasons that these additional pieces appear here (ultimately—for this is what publishers do—to assist the reader in knowing what to think), the overall collection has gained a subtitle. Parable will do, as long as the reader is not confused by the parabola, the curve between points off and on an imagined line, and sees the term instead as the dictionary does, as "a simple story."

It turns out there was all along a story that had only partly arrived that long ago spring and which in any case was indeed more overtly a techno-parable than the others here, and thus, now done, would fit nicely with an essay and a half of more heft and spine where it might not have fit with the more evanescent tales and meditations.

These almost-essays are not, I think, meditations in the sense the others are, though they admit from time to time of the kind of murmur and longing that characterize the others, qualities which an anonymous publisher's reader (whom I have to thank here) links with Asian aesthetics by virtue of their "coherence through dynamic tension."

Perhaps it is something like this that I had in mind, rather than a cheap play on the traditional essayist's term "reflections," in calling the less-overtly narrative pieces here refractions in the newly gained subtitle. To be sure the turn from reflection to refraction suggests a larger (and more dearly purchased) play of meanings. Refraction, according to AHD, is the "turning or bending of any wave . . . when it passes from one medium into another of different density," or again "the apparent change in position of celestial objects caused by the bending of light rays entering Earth's atmosphere," and still once more, "the ability of the eye to bend light so that an image is focused on the retina."

I hope it is not disingenuous to suggest that, in my own life at least, the most fruitful meditations involve such turning and bending, the moves from one or another light or density. It seems right for a collection coming forth from a press located in Albany, so near to Niskayuna (Watervliet), the Shakers' first home, to think, in the words of their well-known hymn, that "to turn, turn will be our delight, 'Til by turning, turning we come round right."

Any such turnings, whether seen as meditations or refractions, upset equilibrium and comfort—driving one toward, one hopes, rather than away from the coming round right. This collection, including its last pieces, is premised on a belief that reader and author are here engaged with an attempt to make sense of our actual existence and especially the stories we tell about it (the moral tales) and the ways we explain it to ourselves (the—likewise moral—meditations). I trust that most readers will come to understand, as a second anonymous reviewer (likewise now thanked here) expresses it, "the writer is in the same position with his book that his characters are in with respect to their conversations."

In this case that position is abiding and biding, and for all their seeming certainty, I earnestly hope that the latterday additions here show themselves as such.

New Hamburg, New York

ACKNOWLEDGMENTS

I did not want to break the flow of these texts with scholarly citation (except in the few instances where citation served historical
and typographic purpose, especially in the latterday essays).
There are page numbers for each citation and, since no one author is
cited twice, it works out fairly well. Works cited follow at the end of
the collection.

Two of the stories, "Storm Tossed" and "White Moths," previ-
ously appeared in print in *The Iowa Review* and are reprinted here
with gratefulness to the editors and by permission. As already noted
above, the latter story emerges from my web-fiction, *Twelve Blue*, co-
published by *Postmodern Culture* and Eastgate Systems, in 1996 and
1997 respectively, for which I retained copyright.

The two essays among the last pieces had other lives as well.
"The Persistence of the Ordinary" began its life as a much briefer talk
given as an invited panelist for the "Innovations in Humanities Education" panel sponsored by the American Association of Higher Education at the MLA in December 1999, where it baffled all but a
small but devoted claque of my readers. It has continued to baffle au-
diences at several universities as it grew like Topsy.

"Afterward: The Future of Fiction and Other Large Phrases" extends remarks I offered by way of introduction to a keynote hypertext fiction reading for the NeXT 1.0 (New Extensions of Technology) conference sponsored by the Media and Communications of
Karlstad University (Sweden) in April 2000. The lowercase letter in

Acknowledgments

the conference title is not a typo but rather an instance of the e-virus effecting e-very discourse.

I would like to thank James Peltz, a university editor with courage, grace, and vision; Dana Foote, a designer with an eye for light; and my research assistant Keri Bertino, who fell into this mess and pulled me out and through it.

Finally, like Alicia, Carolyn always believed in these and all my stories even when I did not, while Eamon and Jeremiah are always the boys whom I continue telling stories to, at least in my heart. I love them each for what they tell me and for their turnings.

MORAL TALES
AND MEDITATIONS

Colossus

In person he is the same man you have seen on television. Words, however, do not serve to describe him. You might be tempted to call him vast, monumental, even mountainous, but already language would have transported you into the ironic. He himself—charmingly, what strikes you first is his charm—is willing to suggest alternatives.

"Perhaps planetary," he offers, "or mighty. Immense, monstrous, mammoth, prodigious . . . "

It is tempting to banter with him as he reclines on the sofa, his flesh covering it over like a wave.

Stupendous, you might say. Titanic, elephantine, gargantuan.

"Jumbo, as in the jet," he bethinks himself, chuckling more delicately than you might have imagined. His jowls quiver in a way that can only be described as fragile.

Jumbo is the correct word, you realize, although not like the jet. He is rather like a jumbo egg, but imagined from within. You have the sense that who he is floats in a vitreous cloud of albumen, his soul or self, or whatever it is that he shares with the rest of us, now swallowed up, tucked away and floating there like a homunculus.

The inadvertent pun is unfortunate (I swear to you that in his presence you do not feel irony, although the fact is you neither feel tragedy nor comedy either) but apt. Looking in his cloudy grey eyes you have the sense that he has swallowed up himself, autophagous and blasphemous, devouring and devoured by the sacrament of himself.

"You've studied the classics," he says. "There are few people who could use the word, *autophagous*."

"You know its meaning?" I ask.

"Self-devouring," he says instantly. "I am like Shakespeare in that much: 'small Latin, lesse Greek'."

He refers, of course, to the perhaps apocryphal description of Shakespeare's education. It is pleasant to talk with him. You recite some lines as if to reciprocate the allusion:

"He doth stride the narrow world and we petty men walk under his huge legs, and peep about to find ourselves dishonourable graves . . . "

He locates the quotation immediately. "Julius Caesar," he says. "But do you know of whom it is said?"

You find yourself thinking that such a man deserves a better age, if not Shakespeare at least Milton who, according to Dr. Johnson, "could cut Colossus from a rock though he couldn't carve a cherry pit." Instead he is a creature whose definition falls to the descendents of Barnum, whether Action News or Jabba the Hut.

You have almost helplessly veered into the ironic. This is a genuine man, a human being, and we lack a language to account him. Not even the worldwide web, forgive my sustaining the ironic tone, can contain such a presence.

"It is the only sutra of our age," he says, "the whole of our meditation. We might spend our lives in a fruitless quest to root irony from our souls. Irony is the faceless void which paradoxically we find ourselves everywhere facing."

When he talks beyond a simple phrase or sentence his voice grows watery and gravid, as if his lungs are caught in the weight of his own expression. He clears his throat and coughs in a rumble. The floor creaks. This is no exaggeration.

"Don't worry," he says through the cough, his voice still thick, "the joists below are reinforced. Everything is. It is necessary."

The simple dignity of the phrase is likewise impossible to convey without irony. Yet you must take my word that he spoke it with great presence and guilelessly, giving each word equal weight yet not weighing them down with self-importance or undue sadness and not unaware of the comic aspect of the comment. It was straightforward for him. Necessary.

You think of the particular Buddha of abundance who is often portrayed reclining while children tumble upon him like a mountain.

This allusion he cannot place. "I am afraid I haven't really studied the Eastern religions," he says.

"But—"

"You wonder about the Silk Flower Sutra," he says.

It is a lovely phrase. You acknowledge that, yes, you wondered at his earlier use of the word *sutra* to describe his thoughts on the rooting out of irony, given this recent claim that he hadn't studied the East.

"We are all Buddhists now," he says. "It is the effect of the media. Nothing is real, not CNN, all-talk radio, or virtual reality. The veil of illusion is within our sight."

Now you wonder if he has introduced irony.

"Can we turn—"

He interrupts again, this time consciously ironic, a twinkle in the clouds of grey. "To weightier matters?"

You share his laughter. Eventually, however, it turns to an uncomfortable wheezing and your own laughter dwindles.

"Can I . . . ?" You do not know what you can offer to do.

He gestures with what looks like a high-tech pipe of some sort but which you almost instantly recognize as a bronchial inhaler. As he suckles on this medication his face seems wise behind the plastic apparatus. The wheezing eases.

"For me," he says in a whisper, "every action has its implications. Even laughter."

On television he has disclosed that he must plan his most intimate ablutions hours in advance.

"Ablutions," he repeats the word. "You'll excuse me for saying this, but it really is remarkable how you use language. Do you find that it is generally difficult to make yourself understood?"

"With some people, yes," I say.

You are aware that for some time now you have had the sense that his fleshiness has receded from you in favor of some essence, a kernel. He seems more directly before you. Not necessarily nearer or in closer focus but more present.

"Like the pit of an apricot?" he wonders.

It is a curiously apt figure for the sensation. The sense of sweetness and smooth stone mixed with a faintly cyanotic aspect. All this flesh seems hyper-mortal, an exhalation of mortality, its span and surface, its mass and substance. Its literal embodiment.

We sit in silence for awhile. His girlfriend enters. She touches his massive arm in a passing gesture of great refinement, not a pat or a caress exactly but a touch, the intentional gentleness which is the lost sense of the word touch. She is, as has been reported, quite thin and not unattractive. Her hand, in the touch, is olive against the smooth, ivory skin of him. It is clear she loves him. What has not generally been reported is that she has been with him for some time and has not, as some have insinuated, merely latched on for the publicity.

"What everyone wants to know," he says, "is do we make love and, if so, in what fashion. That question followed, of course, by the other one which briefly showed itself between us awhile ago: how does a human being grow to this expanse?"

"I am not everyone," you say.

His girlfriend studies you with wary eyes. This is disconcerting since you realize you in some sense intended your response to have a favorable impression on her. It is as if, unconsciously, you vie with him for her affection. As if you cannot reconcile yourself with think-

ing that a woman so slight could favor such a massive man over someone like you, no matter your own infirmities and awkwardness. Her
gaze embarrasses you on this account.

"Sit down, my dear," he says, "Please do join us."

His voice carries a note of kindness and yet you realize—gratefully really—that the invitation is uttered in such a way that he ac-
knowledges your discomfort and attempts however subtly to convey
it to her. Yet you know she sees your unease on her own. You see it in
the way she breaks off her gaze from you, doing so in two perceptible stages with hardly a ripple of time between them, yet articulated
enough that you cannot fail to feel them: first seeing you once again
briefly, and then releasing.

For a moment you are afraid that he will say something consciously oracular.

"You mean something like 'we are all everyone'?" he asks, then
laughs. "Actually I considered it. There was a moment of tension."

"I'm sorry," his girlfriend says, addressing it to you not him. She
is sincere as are you in reply.

"No, forgive me," you say. "It was a bit oracular of me. I'm
sorry."

"Forgiven," she says. Her smile is a touch.

"It is natural enough," he says, "a nine-hundred-pound man is
a curiosity to say the least. How does he defecate? Can he perform
coitus? How much time is required to so unnaturally expand one's
corpus? Does he need a special chair or door or bed?"

"Children ask him the best questions," his girlfriend says.
"They swarm over him, really, and make him laugh. It's like king of
the mountain."

"Or your Buddha," he says to me.

I cannot emphasize too much how unironic was this young couple. How cheerful. How suited to each other despite the disparities
one might notice. Yet already I fear that the mere notice of these facts

in itself leads to a kind of implication. A tone seems understood the more you say none is intended.

"We are happy that you are here," she says to you as she leaves the room, touching your arm.

It is strange, you think, but I expected that he would be eating all the while. I thought when his girlfriend came in that they might offer refreshments. You are, you realize, thirsty.

"That actor who was paralyzed goes to great lengths in interviews on television to make it clear that he and his wife make love," he says. "I saw an interview with the both of them and she was no more discrete than he. It was, I suppose, meant to be touching and frank, somehow with-it, but there was something irrevocably sad about their insistence. Smarmy really."

You know, even by this second time, the rhythm of what follows an extended statement from him, the watery aspect, the weight of lung, the cough, and then the resumption of the state of things. It is like weather. You begin to understand that these rhythms can become a consolation for he who suffers them and you think, in some inexplicable sense, that the same rhythm must have governed his expansion to this state. You realize, joyously, that his anecdote has managed to suggest the answers to the most frequently asked of the commonly asked questions. He is an admirable conversationalist, a colossus of conversation.

No wonder he is in such demand in the media and such a success on television. For where talk is cheap he insists by the pure bulk of his presence, forcing the transcendence of image into language.

"There is of course the third most frequent question," he says, "But the one least often voiced."

You nod in a somewhat too eager gesture of assurance. You understand, you understand, your nodding is meant to insist. It seems ineffably sad that he should have to express what is understood in almost any encounter.

"Will you die soon? for instance," he says, "Will this kill you? How does anyone live like this? How long can you live like this?"

"That makes you the token of our shared mortality," I say. "You embody the questions that are the very substance of our fears. Your size itself makes them evident, constant, inescapable. The interviewer in fact wishes to know these same things for and of himself, but you offer him the occasion."

A feeling of unutterable joy accompanies this recognition. The very speaking of it buoys you and makes you feel more alive and confident. He responds to your enthusiasm, clapping his enormous hands together in a single delighted slap like thunder in a clear sky. In this gesture of applause his arms have moved through space with an extraordinary grace and the lightness of cumulus.

Your thirst has grown almost unbearable with the excitement of your insight and your consequent short speech about it. You wonder why you are not comfortable just saying as much to him, why you are sensitive about bringing up the question of ingesting something, even water.

I'm rather thirsty, you should say.

You realize that your recognition of shared mortality was not sufficient to inoculate yourself against certain fears. Despite your realization you cannot clear your mind of misgivings about the bulk of him and its relation to death. It is as if you fear that even a mention of water might drown this leviathan.

"What shall we talk about then?" he asks.

"Next?" you ask. "Or do you mean to suggest that we haven't really begun to talk about anything of consequence?"

"Can I get you something to drink?" he asks.

You fight against a swell of irony and literalness, as if silently reciting the Silk Flower Sutra to yourself. *Can* he get you something to drink, indeed. You wonder what efforts are required for a simple errand, then realize, stupidly, that there may be beverages quite easily within reach. And in fact there is a large bottle of Evian on a nearby side table only slightly out of sight. There are glasses with it and flexible straws, although not the kind they have in hospitals but rather the novelties of the sort popular with children where the straw loops on itself in a spiral.

"About the reinforced floor joists," he says, "it may strike you that I am the size and weight of a grand piano."

"The floorboards and joists of even new houses—one should say *especially* new houses—are apt to squeak with any weight on them," you retort.

In this exchange you have both moved beyond irony to absurdity, which unlike irony is a form of solitude. Irony is social, an understood wink, a raised glass, a shared sense of exemption. Absurdity is a grand piano.

You would speak this phrase but you fear turning it to merely an ironic bon mot, a raised glass between you. You take the opportunity to rise and cross the room and pour yourself a glass of the Evian, which you drink ravenously, if it is possible to use such a word for thirst. Do ravens gorge on water? you wonder.

Outside the window a television crew emerges from their white van, the camera person, it is a woman, harnessing herself to the gear, raising the black camera on her shoulder. The reporter studies a scrap of script and nervously brushes over the surface of his gelled hair feeling for strays. He is like some mantis, his perfect face rose-orange and from this distance unblemished. The technician turns a white crank, which slowly erects the satellite dish from its nest on the roof of the van. You realize that your time here is coming to a close.

"There is no hurry," he says quietly. "I am not going anywhere. After all that is the story of who I am, isn't it? That I cannot move but eventually will have to be wheeled from here by assembled volunteers, celebrities, EMTs, and firemen."

Aghast, you realize you have not offered him water. You indicate the offer with a lifted glass and he nods gratefully. You pour the Evian and he swallows it in a single draught, though not a gulp, that word too harsh for the elegance with which he takes the drink in. The doorbell rings harshly elsewhere in the apartment and after a delay you hear the girlfriend's footsteps as she goes to the door to greet the television crew. You can tell from the sound of her footsteps that she is barefoot and you cannot recall if she was likewise barefoot when you had your exchange with her. Hearing her barefoot at a distance seems intimate and erotic. Both of you listen to her in the distance. You pour him more water. This glass he sips at more delicately, the cylinder fragile in his large paw.

"David, the waste . . . " I say, using his name familiarly, though not forming the question beyond this.

"I know, I know . . . professor," he says. It is charming, he wishes to reciprocate the familiarity and the short pause between his assurances and the title of address indicates his care in this. He could have called you Mr SuchandSo but that would not have given full value to your own minor celebrity. Yet there is no term in our age for a novelist and essayist. Wordsmith might have done once, but never really came into common use. Instead he reaches into your own biography and plucks a minor honorific from the inadequate sinecure of your visiting professorship.

Your disquietude about your own status and success turns you consciously ironic, self-sorry, and somewhat glum. You have lost your mood of insight and optimism, your insistence on unironic value. It is a relief, a deliverance, that the television crew awaits in the hallway. It promises to keep you from spiralling down into your own fears of disappearance. Yet David shows no sign of releasing you.

"But think," he says jovially, "what exactly is the waste here? I am not a rich man nor am I deeply in debt. Have I used more resources than another man in my situation? People, simple people for instance, manage somehow to buy cabin cruisers and other boats of such a size and expanse that they require a strong truck to tow them and half a city block to park in. Think of the enormous expenditures required for a half dozen jaunts on the water in a single summer."

This utterance extends him almost beyond what you can bear. By the end of it he is weighed down by the weather of his being and he is wheezing deeply. He drains the water glass again but when you move toward the Evian to refresh it he waves you off. Though he fingers the inhaler, he does not as yet breathe it in again. You wait in silence with him, listening to the distant chitchat of his girlfriend and the television crew. For the first time street sounds enter in, automobile noise and an airplane on its final approach high above the city streets. Somewhere there is a windchime faintly.

"Or is it the waste of decorum that people mean?" he resumes. "That is almost surely it, isn't it? Someone expands beyond the bounds of what seems right, like a party balloon taken beyond the point where its illustration reads correctly. The clown's face grows horrifically out of proportion and the letters of the greeting amplify beyond words into a distorted banner, without definition, granular and garish, almost unreadable."

This time he breathes deeply into the bronchial inhaler. In the room there is surprisingly a smell of flowers rather than anything mint or menthol. Spicy flowers, to be sure, violets or something like that.

"They could mean," you say. "I might have meant," you correct yourself. It is you who brought up the question of waste and you do not want to seem disingenuous.

"I might have meant the waste of being. The question of what you are called to."

The inhaler makes him seem vaguely like the elephant Buddha, a figure of great joy and sadness alike. He studies you from beyond this comic plastic apparatus and then heaves an enormous sigh which seems to fill the room with fragrance. He asks a surprising question.

"Would you be among the men who carry me out?" he asks. It is a sincere question, without irony, a service, a boon, a philanthropy.

"Of course," I say. I do not ask whether he means when the celebrities and others come at first to bring him wheezing and breathless to the emergency room or whether he means that more peaceful, final carrying.

"Of course," I say again, "I feel that we have touched each other."

She is in the room then, barefoot and in a long elaborately printed cotton skirt, smelling faintly of patchouli.

"The people are here," she said simply, smiling to him and to me in turn.

Had she been wearing the long skirt then, before? It was an Indian print, mandalic and herbaceous, bright colors on a black field highlighted in gold.

"I wish . . . " you say but do not finish.

You wonder what exactly you wish. To stay with them for the rest of this afternoon? For days on end? Forever? Unlikely. Before long there would be the imperative of the body, expanses of flesh, dank exhalations, cooking smells, sharp words between them, loud televisions contending from opposite rooms. Nothing more or less troubling than ordinary lives really, but ordinary lives nonetheless.

"I do also," he says.

The television crew is already stirring. The technician pushes past the girlfriend—excusing himself but pushing nonetheless—dragging a quilted aluminum valise, a lasso of black cable over his shoulder. The camera person waits behind the girlfriend. She is not so pushy as the technician but impatient nonetheless. She swings the eyepiece back and forth before her eye, testing. Only the reporter is not crowding in. He waits for his moment somewhere out of sight.

Reflection

By the time the woman across the street waved gleefully at him from the passenger seat of her lover's cream-colored touring car as they jauntily passed by on the street in front of her former house not two weeks after she had run away from her grumpy husband who still lived there with his frightening dog and his samurai headband, Jeremy knew everything had changed.

Lynn had told him this exactly in email from the coast. It was the same day that he discovered again, while looking in the mirror, how much she loved him (and he her).

They didn't know the woman across the street and certainly never conceived that her lover, should she have one, would be anyone but Harold, the gentle dumpling of a man, an illustrator in the city, who lived next to her in the former firehouse.

They had seen them walk together sometimes and sometimes smile across their neighboring lawns, Harold sheepish as a sheepdog and sometimes clearly blushing even from the distance of their kitchen window across the street. The woman across the street was vaguely girlish, though broad in the beam of her gardening overalls or her New Englandish denim skirts.

Yet she radiated sensuality, Lynn said, and Jeremy could see it in the slow swing of her hips or her bare toes in the newly mown grass, the toenails painted the rose color of Lynn's lips.

The whole town walked at night, not unlike an ant colony. First the commuters returned from the city in their German cars before or after the few wives who still did not work came back in their Swedish cars from picking up take-out food or hauling kids from soccer. Then there was an hour of silence on the streets in which cats skulked and mourning doves mourned from where they perched on the cable television wires. Before long, doors opened and the adults walked in pairs or shot forth earnestly pumping in power-walk singularity while, doubtlessly claiming the press of homework, the kids mostly stayed behind and watched television or talked on the phones.

Or so Jeremy thought. He knew nothing. They were not part of this town despite the fact that walkers nodded to them and on occasion someone would say something neighborly over the stockade fence.

"Smells good," a man called to Jeremy when he was grilling Thai-style salmon on the charcoal grill.

"It's the coconut milk," Jeremy said and the man nodded and went on walking.

The woman across the street almost never said anything to anyone yet she almost always comported herself with a smile of inner serenity as if she heard fugues in the buzz of the electric lawnmower, saw visions of the Virgin in the lily bed on her side lawn. Sometimes, before she left him, she and her husband walked with the others in the evening, she smiling serenely, he straining against the power of their frightening dog, the leash out before him like a supine kite.

Sometimes Lynn and Jeremy walked but it was as if they were invisible—exotic and transparent tropical fish in a pool of trim and earnest trout. On occasion the woman across the street and Harold would disappear into the shadows behind his house or hers, she leading him inward with the gesture of an outstretched hand or with a gardening basket cradled in her arm like an infant. They gave the air of horticulturalists sharing a promenade, although Harold seemed to walk on tip-toe, his eyes blazing with delight. Often they emerged with cut flowers, although once Harold strode out before her, a boy scout smile on his face and a stack of fallen branches in his out-stretched arms.

Harold was probably gay, although he lived alone and only had friends in for his annual Halloween party and New Year's eve. On summer weekends Harold's brother and his wife would sometimes visit with their lunky adolescent twins, a boy and a girl of great beauty and athletic stature. The twins would join the town in walking after dinner while Harold and his brother played catch on the street with an old football, Harold grabbing the thing out of the air as if it were a loaf of Tuscan bread, his brother tossing the football with great gentleness and evident love for his doughy brother, the ball spiralling like a Buck Rogers spaceship.

Once the ball came over the stockade fence and Jeremy threw it back, too hard, to Harold, the ends wobbling in what the football commentators called a wounded duck. Harold's eyes widened as the ball wobbled at him but then he made a graceful and thoroughly athletic leap and snagged it from the air, grinning to his brother as he reined it in.

"Sorry," Jeremy said.

Another time the grumpy husband joined Harold and his brother for a round of playing catch, for some reason insisting on bending down into a ready position with his knuckles against the pavement and then weaving his way downfield, as it were, along the street in a version of a receiver's pattern. Harold's brother hit him mid-belly with a deadly spiral. It stuck there like silly putty and the grumpy husband ran a few more steps down the street before lofting the ball back over Harold's head.

In the game of catch his samurai headband made more sense to the extent it seemed like a sweatband above his grey sweatsuit. When he cut roses, trimmed the hedges, or walked his frightening dog, the headband seemed a curious affectation, like the ribbon across the head of a china doll. It was something you would ask him about, or probably already know about, if you were his friend. There was nothing a neighbor could do but wonder.

Life was alright there. Along the ridge behind the town there was a county park which wasn't terribly frequented and into which the walkers sometimes took a longer loop on breezy nights when the mosquitoes weren't about. Inside the park was an old burial ground from the late seventeen hundreds where once there had been a Dutch church. Off the path along the ridge there was a cave with dull petroglyphs made by prehistoric Indians. It was a good town to live in.

Some people there were famous and at least one, a man who had a driver bring him home from the city while he sat in the passenger seat of the white Lincoln and read papers or talked on the cellular phone, was very rich. Jeremy himself had been briefly famous. A market analyst, he knew more than anyone on earth about the semi-conductor industry in the Republic of Ireland and, at a time when the president was making a state visit to Ireland, he had appeared on a late-night news program as an expert. His talking head appeared in an oversized box before the news anchor in his New York studio and when the anchor asked Jeremy a question, Jeremy's head replaced the anchor's, taking over the full screen on the broadcast for a moment. Lynn had taped it at the end of another tape they had in which a British nun with wonderful buck teeth discussed a painting of Watteau. The nun had since become quite famous and a regular feature on public television. Lynn and Jeremy could never find a blank tape when they needed one.

Jeremy's fame, of course, did not last, nor did he wish it to. He was embarrassed in fact when a neighbor woman of their acquaintance (she was the landscape painter with a studio that used to be a funeral parlor when the town had one) said something about it as she walked past the stockade fence one morning while Jeremy was sitting in the yard.

"I didn't know you were famous," she said cheerily.

He was flustered at first, he hadn't expected anyone to talk to him. He was drinking coffee and thinking about semi-conductors in Ireland. Her face appeared among the tall sunflowers that made a stalky hedgerow along the stockade fence. The people who walked in the morning were mostly women and a few retired men. In Jeremy's business he did not need to go into an office and in warm weather he often had coffee in the yard.

"Pardon?" he asked. But then he processed what she had said and interrupted before she had to repeat such an embarrassing thing.

"Oh that," he said, "I'm not."

He felt himself blushing. Her face was rather lovely midst the sunflowers though she wasn't anyone who would otherwise interest him.

He tried to laugh modestly.

"Television makes us all famous for awhile," he said, "Your paintings last."

It was too intimate a thing to say to a neighbor woman walking.

"How nice of you," she said and went on. Jeremy noticed how soft her body was in her aqua jogging suit, how full her breasts even beneath the blousy nylon. Her face had been like a flower.

He wondered if this was how Harold felt when the woman next door talked to him.

It had been an exasperating thing to be on television. It filled him with excitements and a vague sense of dissatisfaction. There was very little chance that television would ever again require an Irish market analyst, nor did he desire ever to be on television again. The experience was disconcerting in the extreme. He had driven to the nearby city and gone to the studios of the network affiliate as the news producer in the New York office of the network had instructed. The local station was a one-story brick building like a doctor's office and a receptionist sat inside a glass wall next to the small reception room. They didn't seem to expect him and the studio was closed off with thick glass like the after-hours reception desk at a Motel 6. She rang someone while Jeremy sat in a plastic chair and read the program guide for the network, a glossy pamphlet with pictures of far-off sit-com stars and the equally far-off network news anchor who would soon question him.

When they finally let him inside the studio, a soft-spoken and genial technician led him to the newsroom and offered him a seat on the set. The technician was very tired and yawning, a paper coffee cup in his hand. Across the set the evening news was half finished and about to go to a commercial before returning for sports. Jeremy was surprised that they had just put him out there on this linoleum floor right in the middle of the broadcast and not ten feet away from the local news anchors. The woman anchor smiled at him during the commercial, her face was stiff with makeup and very geishalike in the light of the set, although on the studio monitors it looked as soft as it did on television at home. The anchorman and the sportscaster paid him no mind. The weatherman was somewhere outside the building in a parking lot where he broadcast the weather within the elements, which this night were bright stars and gentle breezes.

The newscast continued as the technician wired Jeremy with a microphone and an earphone that crawled up the back of his head like a worm. Another technician very carefully moved a bookcase filled with cardboard books of the kind they use on furniture store sales floors to a spot behind Jeremy. The bookcase had a small brass lamp and a plastic plant somehow glued down to it. Even so the effect was disconcerting as the technician tipped the bookcase into place without spilling plant or lamp. Jeremy watched it all on a studio monitor set to the left of where he sat.

"You won't be able to watch there when we're on the air with New York," the technician whispered.

"Okay," Jeremy said too loudly. The sportscaster's eyebrows lifted as he read the copy from the teleprompter before him.

The technician ignored it all. "You'll just look into the camera in front of you," the technician whispered, "Pretend that you see him there."

"Who?" Jeremy whispered.

The technician mentioned the name of the network anchor, then thought again and said. "Whoever you please, really," he said, "Just pretend someone is there."

"Like Jesus?" Jeremy whispered.

The technician enjoyed this. "There you go," he whispered and slapped Jeremy gently on the biceps. The sportscaster was clearly annoyed at their whispers.

By now the producer in New York was talking in Jeremy's earphone, telling him that he should stay interested even when he was not on camera since they could come to him without much warning depending on how the anchor felt. "Think of it like a cocktail party," the producer said in the earphone, "You never know who is watching you."

"Or who is fucking your wife," someone said in the earphone and then laughed, followed by a second laugh. It was the technician. He winked to Jeremy. Both technicians were smiling, the one who had set up the bookcase was now behind the camera. Obviously there was a local intercom in the earphone besides the one from the network. The two technicians—the floor manager and the cameraman they were called—joked throughout the broadcast, although they were careful never to disturb Jeremy when he was waiting for directions from the network.

The local news anchors stood and gathered papers after their newscast. The sportscaster wore blue jeans under his news blazer, although you could not see that on television. The woman anchor wore leggings and wool socks and white jogging shoes like a stockbroker at lunch hour. The white socks and jogging shoes were like clouds. They were done for the day and they left the set loudly, laughing together.

"Shut the fuck up," the technician said in the earphone and, Jeremy supposed, the studio as well.

He snuck a look at himself in the floor monitor and, though he did not move his head, the sideways slant of his eyes was obvious. It made him look devious and ferretlike.

"Uh uh uh," the technician chided him. They were counting down to airtime in New York.

"I'm sorry," Jeremy whispered, feeling exuberant at taking the risk to talk in the studio to the technicians.

"Quiet please remote two!" the New York producer chided. The countdown was nearly over. Jeremy's microphone was not local like the floorman's.

During the broadcast he looked into the grey space of the empty and unlit teleprompter above the camera, smiling into this grey void, engaging in facial expressions of acquiescence and concern by turns. The grey was unyielding, reminding him of nothing less than the mothball fleet of federal warships in the river near the city. The host of the program clearly knew nothing of Jeremy's work or his responses to the questions the segment producer had taken Jeremy through for a full hour on the phone the day before. At one point he shifted to the remote guest without warning, asking Jeremy to comment on the Irish president.

"She's quite a handsome woman, wouldn't you say?" the anchor asked.

Jeremy felt a spurt of panic and his mouth went dry. For the first and only time he snuck a look at himself in the studio monitor. He looked prurient and sly.

"That isn't anything I could comment on, Ed. She's much more handsome than a microchip."

It was a good quip he thought and the technicians seemed to agree.

"Attá go," the voice in his earphone said, "Ed would love to plank the president of Ireland."

In the studio monitor the image changed to a two-shot of Ed and Jeremy within the square of a monitor over the host's shoulder.

Jeremy wondered why the image wasn't repeated like a barber's mirror.

"Pssst, Psst! Look at Jesus!" the technician whispered into the earphones. Jeremy turned back to the grey square of silence before him. The program went on without him for awhile then the host asked him for a last comment about the trade deficit with Europe (something Jeremy suspected did not exist) and then they were done. The technicians gathered up the cables, spooling them over their elbows and arms. Jeremy let himself out of the studio and went home to Lynn. He didn't want to watch the tape that night.

He watched it when she went away to California. Unlike him, she didn't travel often. She was visiting a gallery for a show she was curating at her museum. It was an unknown painter, a primitive, suddenly the rage, a sweet old man who all his life painted oversized canvases of a fantastic city of his imagination studding the paint with pieces of broken glass, bottle caps, and the feathers of common birds.

"He asked me to sleep with him," Lynn said on the phone, "Can you imagine? The old letch. He smells like piss and baby powder."

Because Jeremy often travelled to California for the microchip business, he was able to tell her the names of unusual restaurants. She decided on the Chilean restaurant where they served blue potato fritters and roasted chicken.

"You are in my dreams," she said when they hung up, "I'll email you."

He watched the tape then. He had thought he would be foolish looking but the network lighting and the expert cutting to and fro made him seem cosmopolitan and wise, even interesting in the matter of microchips. It was a strange sensation. He tried to think of someone to whom he could show this tape.

The next day the email from Lynn said: Everything has changed. I'll call you later. Love. L.

It troubled him, not because he gave it more import than it warranted (she was probably saying that she might not book the show, or that there were problems, and that her flight would be delayed) but because it made him realize how much he depended on her love.

He sat for a time before the mirror at her vanity in their bedroom. It was a place she never really sat for more than a second, a ritual kabuki theater in which a morning was assessed or an evening's outing received a critical glance, a brush of lavender, the smoothing of an eyebrow. Sometimes when she undressed and came to bed he would see multiples of her body in the triptych of mirrors, her breasts multiplied like ripe fruit, her curves voluptuous and dizzying before she moved out of the space of the reflections, put out the light, and slipped into bed beside him.

He loved her so. In her mirror he saw himself as someone unknown. The face there was unmoving, gentle, open, its eyes glinting with a touch of humor at this situation. He studied himself from a point within himself and thought that the man who gazed back so trustingly at him was someone vulnerable and wise, someone innocent. He longed to cradle the head in the mirror as she did, his hand on the curve of the skull at the thinning temple, moving gently along the contour of the cheek and settling at the jaw. He knew in this instant of impossible longing how much she loved him.

He was standing in their front yard near the crimson Japanese maple when the yellow touring car came around the corner. The car was filled with laughter, the woman from across the street and her new lover both laughing. There was music on the car radio and so much laughter that Jeremy briefly thought the touring car was some sort of advertisement, for a new movie or a restaurant perhaps, someplace named Daisy's or Gatsby.

The woman waved at him and laughed with her lover. Jeremy could hear her say, "That's my neighbor," and he saw the lover hold her more tightly, his arm slung round her shoulder. They went down the street to the circle at the end of the cul de sac and came back up again for one more view of her former house before heading off.

Jeremy looked on as the touring car hurried down the street, made the turn up toward the park and disappeared from view. He was happy that her neighbor, Harold, had not been home to see this. Jeremy wondered whether her samurai husband had seen it all from inside the house. Though he inevitably felt sad for him, he somehow thought it served him right.

He knew nothing, he knew. Two doors down from them a woman had died sometime over the last winter without their knowing. They knew who she was because they had seen her once when they were out walking. The woman pushed a green bottle of oxygen before her in a wire stroller like a narrow shopping cart as she moved slowly up the new ramp whose green treated wood crossed the front of her house like a gash. Silver transparent tubing rose to a hoop which sat like a fallen tiara across the back of her skull and fed sweet air to her nostrils. She reminded Jeremy of his mother. He and Lynn continued down the street while the woman continued up the incline. As far as they could see, there was no one outside there to help her, although they could hear the sounds of television from inside the screen door.

Reflection

The woman or someone had painted a blue handicap symbol within an uneven rectangle of blue paint in the street before her house. The symbol and the lines were thinner than usual street markings and the blue paint was slightly brighter than what one was used to in these markings. The symbol seemed meant as something like an evil eye or the Indian petroglyphs, something to ward off death. The lines were largely straight though they wavered somewhat like the surface of a distant river.

FIRST MEDITATION
At Home with a New Thing

B
ringing technology home seemed a father's task at first. The father of Gregory Houston—the first playmate in my memory with a personality and history, although I recall now only a plump, buffed sort of blondness which I have come to associate with a certain class and wealth—was a doctor. I do not think I met his father, a recollection that would seem absurd on the face of it but fathers were often unseen in those days. Although our neighborhood was small—we lived for my first years in my grandfather's house in a mostly Irish ghetto but along an upscale avenue which Olmstead designed as part of a spoked city of parkways, their grand elms forming leafy tunnels over the roadway—there were evident classes and seniorities and a doctor didn't promenade. In any case Gregory Houston's was the first television on McKinley Parkway and we used to watch Howdy Doody there, in the eponymous city of Buffalo Bob.

Locality was an essential aspect of the way I formed my understanding of media. The programming was clearly fastened to our sense of place, just as my mother would listen to the radio all morning at the kitchen table, moving easily from the Clint Buhlman program, local, to Arthur Godfrey's networked one. They were "favorites of hers," a phrase an eldest son understood with a special grace and locality, a sense of possession. On the television also, besides Buffalo Bob, there were favorites. "Meet the Millers," a daily talk show not much different in kind I realize from the offering of the various "women's" cable networks, showcased Bill and Ethel Miller, his face narrow as a wedge of bundt cake and his bearing as stiff but with a hint of a bemused smile, her voice whiskey rasped and herself dark and ample, she might have been Jewish, although the program had a distinct waspishness (in both senses), which often put my mother off. Still they were neighbors of a sort, broadcasting from their colonial home in Elma, a Buffalo suburb, although we were as unlikely to use the words *colonial* or *suburb* as we were the word *technology*. As Christmas neared there was also a local Santa show, which we somehow knew was local even during the time we maintained our global belief in the global saint, a program with wonderfully literate repartee between an ironic Santa and two thin elves, older men with rubbery clown faces in tights and pointy hats. Even so we made no real distinction between the local Santa and the group of Mousketeers, whom we also endowed with the sense of locality, which I realize now we were supposed to. These shifts, from Buhlman to Godfrey, from Santa to Buffalo Bob to Scotty and Darlene, were, I think now, my first experience of being programmed.

I have a very firm memory (unusual since my memories are mostly a shifting fog) of first hearing about television from Gregory Houston. We were out on McKinley Parkway, it was dusk and winter. I have a sense we were rolling bales of snow into rounded forms for a snowman, a sense also of the yellowy carlights (higher then, almost carriagelike on the arched, round fenders) on the parkway, the sound of slush against tires. He must have said something about television, perhaps even bragged. I am certain the memory snagged there because it was a word I had no place for. Television. I have a similar memory of a faintly populist dismissal, a thin mix of repulsion and something unidentifiable—not so much envy as helplessness—about the Houston's status. A doctor's family. We didn't have a television until a year or more after them.

My father was a wiry, strong man, a steelworker, the kind of man capable of lugging any television home. Like most, ours featured a sturdy mock-mahogany cabinet with hinged doors and lacquered hardware in which the grey eye of the tube reposed. My father tinkered and early on he showed me the thick, cloudy glass of the funnelled posterior of the tube, with its waxy bundle of copper wire: the transformer. A scent of electronics, not smoke exactly but clearly heated, like no other scent and unmistakable since. When it came time for color television my father was able to do what other fathers couldn't, setting the magnets just so to bring the colors into adjustment. By then we had veered from mahogany toward fifties modern cabinetry, with blond wood and burnished hardware.

What one brought home was a shadowbox of sorts in which the world appeared, silvery and slightly out of focus. There was a greyness to the images, which did not appear quite the same elsewhere in the natural world, the grey veering toward both brown and silver. Many commentators have written about the shift from the participatory radio to the more passive television, how programming and technology alike shifted from a sense of audience as multiple to the singular viewer, how certain entertainers were not able to make this shift, in much the same way that vaudeville comics could not easily shift to movies. I suppose that is all true, but the changes which came over us at first were minor shifts of coloration. Not only were we largely unaware of them, but it isn't clear how awareness would have served us.

Kinescope did have a noticeably different color, more of the brown tones as I remember. It signalled the time shift, the fact that what one viewed was not current but somehow held in this brown cloud to be distributed later. After which it disappeared as live programming did. Or so we thought (then later seemed shocked when it sometimes turned out to be true, that a surprising number of television archives lack the earliest years, copied over). It is hard for my children to imagine but we had little sense of taped versus live at first;

rather there were different live events, some which had been ki-nescoped. Even more difficult to understand was how the funda-mental relationship between the world and the television differed. Be-cause of the sparseness of programming at first there was little sense of the television as a receiver, an apparatus picking among signals as we were used to on the radio. Instead the television was more like a camera (my father was a photographer, perhaps Gregory Houston saw the television as a stethoscope or its visual equivalent, the coni-cal whateverscope used for looking into ears), grabbing segments of what we did not then know to call real time. Now the television seems a conduit for images which, otherwise unseen, wash through the at-mosphere. Sometimes it is also an aquarium through which exotic and elusive creatures swim for hours on end, caught up in the distant and several mysteries of their lives, untouchable and immaterial.

Besides having the first television, the Houstons had a swing and slide set long before anyone else had one, and I was attacked by a stray German Shepherd dog there. The dog came out of nowhere, with teeth like the wolf in Hansel and Gretel. I was sprawled on my back on the scuffed ground near the swing set and screaming, terrified but however improbably holding the beast's jaws open with my hands. Somehow, from a full block away and like an episode of Su-perman in which Clark Kent hears a disturbance and must slip away from Lois and Jimmy, my father heard me screaming and saved me, chasing the dog off. Perhaps Gregory Houston went after him. I do not think that on that day Doctor Houston came out from the house, where he also had his office, although I suddenly recall now that their house was constructed of pocked, honey-colored brick in a neigh-borhood where generally most houses were made of wood. I can also (this is the truth and not a rhetorical gambit) suddenly recall what Doctor Houston looked like, his face, too, a little plump but kindly, with warm dark eyes and a five o'clock shadow beneath fragrant and slightly florid cheeks.

Storm Tossed

In a gesture of solidarity his son went birding with him in the days following the hurricane. There had been reports of storm-tossed birds in the valley, wood stork, skimmer, and sooty tern, each of them carried in the envelope of the storm's eye and delivered here like temporary miracles.

It happened sometimes. In August 1933 a herald petrel was sighted near Ithaca, New York, the first and only sighting of the species outside the south Atlantic. These misplaced birds were really not so much carried as left to fly for their lives in a moving bubble of still air, itself conveyed at the center of a great perturbation, a pocket of calm flying within storm like anyone's inner heart.

He would try not to say such things in his son's presence. His son was, for whatever reason and for however long, in retreat from poetry. Or at least his father's version of same. At college his son had become a sociology major and now he saw all things as forces against which we huddled in groups, like people getting out of the rain await a bus in a glass shelter.

He had wanted his son to study art or poetry or geology or even plumbing, something with substance and mystery. Yet his hopes were so clearly projecting himself on his offspring that he had not said anything. His own degree was an MBA from Wharton and now he was being downsized.

He should have stayed with poetry, he thought.

It had been three days since the hurricane, which had struck further down the coast and headed into the Appalachians. It worked through the mountains like a tractor dragging a thresher of sharp, sustained winds behind it. As the storm muscled through the mountains, microbursts of rain loitered behind like bullies in isolated villages and hollows. A full fifteen inches fell like a biblical plague on a Pennsylvania town that had already lost the members of its high school French club in a terrorist attack on an airplane.

His son had been to twelve European countries by the time he finished high school, all on his own. He thought the world was a neighborhood and now he slipped from the car and grabbed his field glasses and his Walkman to head off into thin woods with his father in search of alien birds.

His father would not say anything about the Walkman although he did not think you could bird while your ears were filled with other music. It was still grey and humid after the storm, even more humid when the sun found wispy patches of blue.

The storm was named Franz, after Kafka he supposed although these days the baptism of tropical storms was generated from a database of unusual names of alternating genders. Whether Franz or Zelda or Cabot, they had no history and were connected to nothing but themselves.

It was similar for the wind-blown birds. Once the storm lost its strength, any surviving bird would drop, exhausted and lost, from the popped bubble of the storm's center and land without history and connection. Out of instinct they would head for rest and water and with luck would find their way to rivers and from rivers to the sea and so southward home.

Like his son, he too carried electronic equipment on his waist. The Garmin GPS 40 personal global positioning navigator was not much bigger than a Walkman but it listened constantly to eight different satellites, offered up to two hundred and fifty waypoints, and ran on AA batteries. It was possible to store thirty reversible routes of up to twenty points each, though the math didn't add up and he wasn't certain that his life held thirty destinations in any case.

"So you'll be able to write up exactly where the birds are?" his son had asked when they packed the GPS in the car.

"Birds aren't current events," he said smartly. "They'll be long gone from where we find them by the time we report them."

He had not followed with the pun that had come to mind. How, borne on by winds, birds were paradoxically nothing less than the events of wind currents. For his own part his son had seemed to forego the natural question about why they would take along the GPS otherwise.

Surely it wasn't possible to be lost in the local?

They were learning pretty well what not to say to each other. Which made space for all they really did have to say to each other now that the storm of adolescence had passed and they could try to understand what it meant to be together as men of different ages.

Thus, for instance, he did not say aloud what came to mind as they first moved up the brambly slope toward the ridge above the creek where they would bird. He was thinking of something he had heard a fat man say on Christian television while, sleepless, he sat up all night and zapped through channels with the remote control.

"Our happiness comes through the mercy of God," the fat man said and mopped his brow. He was lecturing people—he supposed the word was preaching—from a stage set made to look like the lobby of a convention hotel, although it was a little too luridly done up in red satin wallcovering and gilt furniture to really be one.

"Our happiness comes through the mercy of God," the man repeated it, more softly now.

As he sat watching this lonesome spectacle, he suddenly realized he had no sense of what was meant by mercy. It was not simply this religious usage; he didn't know what the word meant at all. He knew he had been taught it at some time, could even recall other instances of its usage: he begged mercy, mercy me, have mercy, mercy sakes.

For a moment as they trod up the ridge behind each other—in his case more short of breath than he wished—he considered asking his son whether they taught mercy anymore. Or at least if he knew what it meant.

The urge to ask the question was prompted by what he sometimes felt as a refreshing awareness that his son knew things that he did not. The decision not to ask the question was prompted by the fact that the question could seem laden, even vaguely accusatory.

They lived in an upscale suburb, at least for now, and their cable system was unique, not linked to any conglomerate. One channel showed the great movies of Godard, Bertolucci, Varda, and Truffaut. Another shifted after midnight to a Tibetan Buddhist network with softspoken and laughing men and bright orange colors. There was even a channel which ran nationwide classified ads from the *Wall Street Journal* news service.

It wasn't anything you would want to lose.

Mercy, he knew, was when you could harm someone but did not do so. Torquemada and the SS could show mercy, as could a mugger. By extension a storm could also metaphorically show mercy. A father could mercifully forego certain questions, as could a son.

He listened very closely but there was no leakage from the earphones of his son's Walkman. It was an atavistic instinct. When his son was younger the tinny noise from the earphone would upset him unreasonably. For some years he patrolled all silences and was ever vigilant for escaping noise. They had grown beyond such regulation.

He thumbed the controls of the GPS and saw the reassuring display of the test screen. Thoughts of regulation reminded him of a curiously poetical sentence from the specifications for the GPS. The sentence said that the GPS was "subject to accuracy degradation to

100m 2DRMS under the United States Department of Defense imposed Selective Availability Program."

It stayed with him. He had easily memorized the sentence. He was known in his company for this ability to get and retain arcane information verbatim on just one seeing. It was a valuable ability.

The sentence from the specs seemed to suggest that the GPS was purposely inaccurate under certain conditions. It was meant, he supposed, to keep spies and terrorists from navigating too securely.

It was like the old saying: keeping someone off balance.

They hadn't seen anything much.

"Just the usual suspects," he said to his son.

"What's that?" He fiddled with the earphone.

"I'm sorry, I forgot you couldn't hear."

"I can hear alright," his son said, "I just didn't understand what you were talking about. What usual suspects?"

"It's a movie," he said. "It was a movie. I'm surprised you don't know it. *Casablanca.* With Humphrey Bogard."

"Bog*art,*" his son said. "You said Bogard, D, like Dirk Bogard. It still is," he said, "It's still a so-called cult favorite. I just didn't hear you."

"I said the usual suspects meaning just common birds. Beautiful nonetheless."

They had disturbed the silence and now it would have to settle again, like when your hand stirred a pool and then you had to wait for the surface to turn still again. This was why one did not ordinarily carry on conversations while birding. You could hear the warning screeches as the whole forest signalled following an exchange between you.

The silence, in fact, was as much a noise as a silence. There was a silence which was the noise of ordinariness. It was what one summoned.

They had come into a stand of black walnut trees, which at the full of summer was still and cool, a dark canopy through which little light penetrated. Now it was thinning with the onset of autumn in late September. Even the grey light streamed through. Black walnut like willow yellowed early.

It felt like cancer, the wasting away of shade and protection, the thinning of hope.

"I love you," he wanted to say to his grown son but didn't.

From time to time his son would stop, as if impossibly hearing some new sound through the music that filled his ears. He would raise his field glasses and search the trees or the slope down to the creek below and then, just as quickly and without comment, continue softly onward.

He had taught him to walk quietly. He was proud of this. A son who could walk quietly could see more.

He wondered what the music was on the Walkman. Not *The Moonlight Sonata* he supposed.

He smiled to himself. It wasn't making fun. He liked the music his son played on the car tape player sometimes. Not all of it, of course, but some, although he couldn't recall the names of anything and so wouldn't be able to say anything to him.

He didn't remember everything, of course, had no photographic memory. Merely a knack for certain key pieces of information.

This morning for instance he had been mistaken. There was mail at home from a William Palmer, Orchidist, at a Florida address, with a special offer of a free gift. When they were leaving to go birding he noticed that the return address said Orchardist. The slip had made him inordinately sad.

Orchidist and orchardist. Lost between the flower and the fruit.

You wanted to avoid saying things you would regret: Would we ever have a garden again if we lost this house? How will we pay for your college?

Storm-tossed, we are lost between the flower and the fruit. There was a certain poetry to that. He liked the phrases.

"I like that," his son said. "What's that from? Shakespeare?"

How strange. Had he said it aloud he wondered. He must have.
"Me," he said, "me."

"Cool," his son said.

"I was just thinking," he said but his son had already begun to
say something more. He asked him to repeat it.

"Do you have a golden parachute?" he asked.

How naive and ridiculous, how wonderfully caring. What did
they learn in Sociology? Idle phrases and popular notions.

He met his son's eyes and saw a storm of worry.

"Not really," he said. "Those things are in newspapers and up-
per management. There will be outplacement, of course. Severance.
Counselling. Transitional support."

All these were words, he realized, no different from golden parachute, wood stork, skimmer, or sooty tern.

"Good," his son said.

How could he hear through the music? How could he see
through the gray?

"Wait!" his son hissed and whispered, grabbing his father's arm.
"Look!"

In an uphill swail, turned to a boggy pond in the aftermath of
Franz stood a Great Blue Heron. For every pond a king. One of the
usual suspects, yet always beautiful nonetheless.

Ardea herodias, one of the day herons, subfamily *Ardeinaethe,*
family *Ardeidae,* order *Ciconiiformes.*

Sometimes words showed you where you were as certainly as
the GPS 40, though they too were subject to accuracy degradation.

His son thought it was something. And it was, a temporary
king, storm tossed here as certainly as a sooty tern was elsewhere. He
begged him mercy. It was something, the heron and the son.

The bird studied them with a calm eye then lifted, its wings
flapping haughtily, barking its forlorn, disdainful cry across the sorry

pond and down the ridge, through a gap to the creek which eventually ran to the river.

Nothing was so certain or so linked. Nothing so simple a consolation.

On the way home he tried to say what could be said.

"What were you listening to while we walked?" he asked his son.

"Do you want to hear it?"

"Sure."

It was more surprising than Beethoven. A spoken word book in which an actor recited lines from a story by Chekhov, words so sad and splendid they seemed like music to him. It was a consolation. They drove the rest of the way home in the birding silence, the one that was as much a noise as otherwise.

White Moths

In Ontario her mother knew a girl named Delores Peters (she remembered the name, "sad stones" her mother said, "it means sad stones,") whose father had purchased a carnival ride and brought it home on a trailer hitched to the back of his Mercury. It was a children's whirly wheel, blue metal cars like stumpy little shoes, each with a black rubber safety belt and running around a sheet metal track, the track scuffed and burnished to a dim mirror where the car wheels ran and the children walked getting on and off the ride. The cars were attached to the end of blue steel spider arms with a string of Christmas lights down their centers and ran round and round the polished metal track without spinning or rising, one behind another until you squeezed the handle of the greased upright brake and slowed them to a stop.

It had been a bargain. The man who sold it owed a gambling debt to a gangster from Niagara Falls and he needed to move fast. He showed a picture round the taproom of a tavern in Crystal Beach, itself an amusement-park town. The picture had scalloped edges as if cut with a pinking shears and the image was a little faded in some parts and enameled in others, or so her mother remembered over the years. She knew that Delores Peters kept it even after her father's death.

The tavern parking lot was unpaved and huge, large enough that sometimes tractor-trailer drivers parked there while they had an ale, easily pulling in and out without the need to turn their rigs around. It was a dusty space that left a taste of stone in your mouth when you walked across it.

There hadn't been time to go back home and get the pickup truck when Delores Peters's father bought the whirly ride, and so the man who sold it threw in a ball hitch as part of the bargain and then helped him balance the load of the Mercury using blocks of wood to shim the shot springs. The ride was surprisingly compact on its trailer. The blue cars stored up on end like cupcakes, the curves of their metal pitted and scarred from being painted over where it had been chipped. The spider legs folded like a pocket knife when their bolts were loosened. The engine and the upright brake were bolted to the trailer together with a folding metal stool where the operator could sit to take tickets or run the ride.

The whirly wheel ran on either 220 or 440 power and had a noisy gas-powered generator for when there was no power available, although the generator roared so bad it made the ride unpleasant and sometimes stalled or spewed black smoke, which tended to frighten children. Delores Peters's father paid nine hundred dollars cash for the whirly ride in 1965. He was a farmer and his credit was good and the bank in Crystal Beach lent him the cash on his word. It was something you could take from fair to fair or just set up on its own for children to ride.

There hadn't been time to go home for the pickup, yet he and the stranger went into town to get the loan and the cash. For an awful moment after the stranger drove off Delores Peters's father feared that he had been fooled and his stomach felt hollow from the beer, the sunshine, and the dusty stone parking lot. But the ride ran just right when he got it home and he was able to set it up without much trouble although it hadn't come with any directions but the fading scalloped snapshot.

He tapped power from the 220 line in the barn and ran the thick black rubber cable to a nice shady space not far from the chestnut tree. There weren't neighbors nearby to watch this effort or it might have gathered a crowd. Even so, Delores and her mother watched expectantly and not without some puzzlement. Despite the shade her father sweated, his dark shirt turning moist under the arms and a damp upright swatch forming from his chest to his belly. Sometimes when he caught her eye he smiled.

There was a trick or two to getting the whirly ride set up just right. You had to be plumb level or the spider arms would shudder slightly and groan and the wheels of the blue cars would skid or balk a little by turns. Still he got it going all right.

"Have a ride," he said to Delores when he was done. He gestured toward the blue cars on their track like someone pointing to a dance floor.

"You too, love," he told her mother, "It's strong enough for both of you if you ride across the way from her."

At first it was very strange for them, mother and daughter on opposite arms, one at three o'clock and one at nine. They were not certain whether they were chasing each other or balancing each other, and if chasing each other who came first and who behind. It was also very quiet at first, only the wobbly rubber noise of the greased black wheels along the burnished track and the soft, urgent straining of the electric motor. For awhile they may have wondered if they should have made a noise, whether squealing or laughing brightly, and how exactly they should let him know when they had had enough.

He watched them happily, the ride was running smoothly and it was a bargain really, though he wasn't really one to show a range of emotion. In time he discovered a small basket weave suitcase tucked next to the wheel of the trailer, the straw colored basketry frayed and torn but the latch and hinges still shiny brass. Inside there was a record player with a large speaker and two different records with tunes to play along with the ride. It was a surprise. They waved when they wanted him to stop.

Delores Peters's father wanted her to invite her friends over for a ride on the whirly wheel and she of course wished to do likewise at first. From the beginning there was something wanting in the experience, a curious sense of mixed expectation and loss. It isn't common to go somewhere where there is a single amusement ride unless it is a carousel and then the mirrors and all the llamas, camels, and horses make it seem more than a single ride and the calliope and colors form a spectacle.

The whirly ride was modest and not a spectacle. In the sunshine the uniform blue enamel had the flatness of weathered house paint although it was clearly fresh and even possessed of a little sparkle. The blue cars seemed somehow Dutch, perhaps on account of the delft color, perhaps because their snub shape suggested foreshortened wooden shoes. Still Delores's mother and father did their best to make the occasion something festive for her friends. Her mother made a frosted cake with roses like a birthday and put a jug of lemonade on ice in a washtub, covering the jug with a dishtowel to keep the yellowjackets away. Her father added other records to the two that had come with the record player in the straw-colored suitcase. The song called "How Much Is That Doggie in the Window?" was a favorite with the girls. He also found ways to vary the speed of the cars, manipulating the brake somehow to add variation to the ride and making the girls squeal at first. When that wore off, Delores's mother ripped strips of muslin to make blindfolds for any girl who wanted to ride with them and both her parents helped tie them over the girls' eyes.

Before long of course the girls drifted off to other play but Delores Peters's father did not seem disappointed. The ride was a success in what it brought together and it was beautiful in its way even when it went around without any riders.

Most of the girls had to go home before nightfall, which was still fairly late then in Ontario in August, but for those who could stay the whirly ride was wonderful with its arm of strung lights in bright colors. The music from the record player seemed both distant and near in the summer night, like a dream and a memory at once.

The girls who remained rode dreamily, taking interest in the ride again under these new circumstances. Delores's father adjusted the rheostat that controlled the speed of the cars so that they circled lazily and the hum of the wheels softened to a low buzz like locusts. The ice in the washtub had melted to a dark pool and white moths caught themselves on its surface struggling against wet wings.

As the parents of Delores's friends came to pick up their girls, they would mostly stand there for awhile watching the spider arms of light go round and flash against the chestnut tree, the barn, and their own faces. Her father nodded silently from where he stood by the upright brake, forgoing his seat at the folding stool, keeping watch over the ride in the night. He waved to the girls and their parents as they left.

Finally when everyone was gone, he invited Delores and her mother to ride again at three o'clock and nine. This time they seemed to spin like people holding hands, not chasing each other at all.

Her mother did not know what eventually happened to the ride. She seemed to think that Delores Peters's father had rented it out once or twice to a church fair or a volunteer firemen's picnic over the years. She also felt certain that he pulled it out from the barn and set it up near the chestnut tree from time to time again, one time in particular when they were teens and rode the whirly wheel ironically with their boyfriends, cramped one to a car. Her mother also wasn't sure what had happened to Delores Peters exactly. She knew Delores had married unhappily but then remarried happily. She also knew Delores's father died before her mother, who in fact she thought might be living still in Ontario, since farm wives tended to be hardy.

SECOND MEDITATION
Time Zones

People like the phrase "real time" and thus use it both too frequently and wrongly. Often they think it means immediately:
let's try to do this in real time, a colleague says. How else, one
wonders, knowing, a little sadly, that real time will unwind more
slowly than she imagines now and without the hoped-for allure such
a doughty resolution bespeaks. Real time is as much ennui as energy,
a modern Pascal might have said in a bad translation.

If so, so much the better. Our present enemy (O why must you
garb it in battle metaphors? my imaginary colleague might, rightly,
ask), our present (I begin again—she smiles benignly) is robbed by
a constant nextness, a world in which nothing happens that isn't
shadowed under the regimen of nextness, and so as a result nothing
happens really.

Really everything happens, despite our lights (a cliche which
takes on a new bite before the various screens where we sit) much the
way it always happened before, sun falling successively over Cleve-
land, Dubuque, Fort Collins, and Pinole. To be sure, the sun never
sets on New York New York Las Vegas Nevada ("visit Vegas and Ep-
cot," my son says ironically, "and you've seen the whole world, no
need for Paris or Istanbul"); to be sure sundry real estate executives
and new media airline magnates spend the equivalent of the gross national product of some of the republics they dangle over as they at-

tempt to balloon into constant sunlight (most of them unaware that it was poor Verne, most prescient seer of cyberspace, rather than Michael Wilder, first husband of Liz Taylor, who wrote *Le tour du monde en quatre-vingt jours*). We follow them as if it mattered. "There remain no more geographical discoveries," Peary was said to have said, safely back on the boat. Chances are that you or our balloonists don't know—I didn't either—that if Peary indeed reached the pole, something we'll never know for sure and so ought to settle upon, he was accompanied by Matthew Henson, a black man, and four Eskimos named Coqueeh, Ootah, Eginwah, and Seegloo, at least as it is transcribed. History in this guise is either quiz-show cleverness ("Great Sherpa Guides for $600, Alex.") or ideological bookmarking, a conscious attempt at recollection in lieu of a genuine multicultural genealogy.

"Time is the substance from which I am made," writes Borges, "Time is a river which carries me along, but I am the river; it is a tiger that devours me, but I am the tiger; it is a fire that consumes me, but I am the fire." Billionaire balloonists ride the jetstream (one of them runs an airline and record label named Virgin which is a rhyme for a record label and a computer company named Apple; another heads a real estate empire named Remax, an instance of a wordlike nonword, Haagen Daz for instance, which almost surely, one thinks, stands for something, but unlike real words has none of the substance of time). They dangle in real time, in real words and, in truth, they do dangle rather than fly: sitting in a metal cylinder and watching themselves on television screens, the jet stream on a computer monitor, the Himalayas far below, smoke from the incense of Lhasa reaching past them to the heavens for over a thousand years.

Real time has looped back upon itself (it is hard to avoid the pun on reels). The CNN reporters watching from a Baghdad rooftop at eleven o'clock Eastern Time for expected bombing runs whose schedule is relayed to them and their anchors by the Pentagon must explain the muezzin's morning summons to prayer cranked from the loudspeakers upon the background minaret as well as the headlights of the cars heading home from civil service jobs through the early morning, prebombing light in the real time of the country we are bombing soon.

Perhaps the civil servants go home to tea. In Jafar Panahi's film *Badkonake Sefid* (*The White Balloon*, 1995), there is a shock in seeing the normalcy of the life within inner courtyards. Wives brewing tea, children with goldfish. That was a previous enemy, Iran not Iraq.

It is later than we think on television. And elsewhere.

We don't have time to wait in real time through the night for the bombing runs, thus they cut to them on schedule and loop them in rerun "updates." Real time is flooded with rerun. The schizophrenic is fabled to hear phrases recurring in his head, again and again and again and again, the word *again* maddening him and dulling him at the same instant—"at the same instant" itself a phrase entirely uncongenial for the endless repetition of either neurotic or mediated experience. Real real time (we have to multiply these like the schizophrenics of fables, I sometimes use the phrase "real virtual reality" to differentiate the impoverished current state of what my students imagine in movie resolution) allows respite on account of its dullness and comparative emptiness.

Adolescence is the dullest season. You keep wanting something to happen, sometimes do your part to assure that it does, and still your life creeps on, dully and emptily, in real time. You stand at the window and gaze, at the mirror likewise, and now the computer screen. It is life in wire-frame. The digital clock is an entertainment, always a new story, always the same story.

Sometimes however the suns sets successively. When I was an adolescent I followed successive dawnings of the new year by radio, nested in my covers in the lordly room I shared only with myself, whereas my sisters and brothers had to bunk down like the denizens of Spin and Marty's dude ranch on the Mickey Mouse Club. I recall the radio as a crystal set, although it doesn't make sense in terms of the rest of the story, which involves monitoring separate broadcasts from Chicago, Denver, and Los Angeles. I do recall it had a single earphone of hard plastic, which you screwed into an ear, and a tuning coil of copper wound round a waxy orange cylinder.

One explanation is that what I imagined (even now) as successive broadcasts across the landscape were instead successive reports across the landscape from a local broadcast. This whole notion seems as confusing to me as Genette's "Narrating *n* times what happened once." In any case it makes me think of an endearing, but also for him a profoundly embarrassing story, about one of my brothers who, sitting at the kitchen table while my father tuned in distant stations on his own superheterodyne radio, refused to believe him when he claimed to have summoned Chicago from the tuner. My brother, then a young boy, believed that if it were Chicago, Frank Sinatra would be singing "Chicago Chicago, that toddling town . . ."

Another explanation, of course, is that there were monumentally strong clear channel AM stations in those days, virtual Himalayas of radio frequencies, pouring across the prairie night and Eastern cities. "The geometry of landscape and situation," says J. G. Ballard, "seems to create its own systems of time, the sense of a dynamic element which is cinematising the events of the canvas, translating a posture or ceremony into dynamic terms."

I am not sure about the cinematic elements of my own adolescence, whose most dynamic elements were fantasies. How else explain my retreating to my bed before midnight on New Year's Eve in time to hear Guy Lombardo on my cylcops-eared radio?

"The greatest movie of the twentieth century is the Mona Lisa," Ballard says, "just as the greatest novel is *Gray's Anatomy.*" Sullen and alone and celebratory, the distributed consciousness of an adolescent is enigmatic as smile, certain as cartilege.

Speed of Light

To her son it was like a rosebud, the toothbrush. It came Federal Express from his new girlfriend while he was in town taking a deposition and staying with his mother at the apartment and he opened it right at the breakfast table.

It was light blue and well worn, the bristles splayed like a book that has been damp and dried into a bulge like a paper pineapple fan. At first when it came she thought it was important papers. But then you could see the love in his eyes.

"This is strange, isn't it?" she asked him.

"You mean is it a toothbrush."

"That I can see," his mother said, "I mean is it *only* a toothbrush?"

He didn't answer. It was of course.

"You left your toothbrush behind and she sent it to you by Federal Express," she said.

"No, it's hers, " he said, "You will love her, mom. She's always doing things like this."

She wasn't certain what a thing like this was exactly.

"I've heard of women sending panties and such," she said.

"Don't get current on me, mom," he said and laughed. "Heard where?"

He still buttered toast like a boy, his wrists somehow twisting round themselves and the knife jutting awkwardly back toward his chest as if in one of those experiments where you have to do everything in a mirror.

"I have a sensual life of my own, you know," she said, and then because he was so boyish she let him off. "Also I watch television."

It was always a mistake to let a boy off the hook. He finished the twisted buttering and snorted.

"More of the latter, I suspect," he said.

"Don't be so sure," she said. "I could make you uncomfortable enough with even an innocent story. Your generation didn't invent sex, you know."

There was something strange in all this talk. They were discussing a toothbrush.

She would be damned if she would give him the satisfaction now of asking whether there was a note in the package or how it was that someone could spend twenty dollars or whatever to send a toothbrush across the country.

In any case she wouldn't judge this woman at a distance.

Still it stayed with her all day. He dressed and went downtown, she went out and back on midmorning errands. The toothbrush sat there on the buffet where he had left it at breakfast. He was really rather badly lacking in social graces; apparently it wasn't something they covered in law school. His lack of training was, she supposed, her doing. Conventions weren't important to her in raising either her son or her daughter, but for her son they were even less important. It was her view that boyhood was a state of unconventionality and the idea of mothering was to help him find the grace in that.

He had turned out well, although she could have done without the conventional barrister's blue suit, rep tie, and shiny black shoes. He, however, was a comer and he wanted to be a partner. He buttered himself better than he buttered toast.

She picked up the toothbrush feeling unreasonably surreptitious. There was no one home but her and he had left it out and in any case wouldn't notice whether she put it back with the bristles at ten o'clock as he had left them or turned the handle to seven thus moving the bristles to one.

For a moment she considered going back to bed. The surreptitiousness following so soon upon the evocation of her sensual life had left her feeling in a holiday mood, as naughty as after drinking brandy or aquavit. This morning when she went out the air had been fresh as new sheets and the streets were black licorice and the park was green and dewy and even the yellow taxis shone. The city got storybookish sometimes and her son was in town.

She sniffed the toothbrush like a rosebud.

There was really nothing there: perhaps, you could convince yourself, a faint whiff of minted dentifrice and, with more mental work, the merest sense of enamel and rosy exhalation. All nonsense of course. The gesture here was the thereness. She admired this in the other woman, Jennifer—god help us, once all the world was Jennifers.

There was a faint calcification, a powdery paste of white, in the eyelet of the brush. It was soothing to see a fault. The light blue plastic was pearled and slightly unreal like the cartoon-colored frosting of children's birthday cakes. You couldn't judge much by that. There is a party game where one asks husbands to say what color is their toothbrush, what underwear they are wearing, what underwear their wives wear. It is good for laughs and sometimes strangely sad. How little we see of what we are or do.

God save us from philosophy, she thought, and almost turned on the television. She thought briefly about calling her daughter as well then decided to wait until evening when the rates came down.

Love, she thought, works strangely.

Philosophy was worse than sexual fantasies, you couldn't turn it off as deftly as you could fantasies. Jesus! she thought, can you imagine, "love works strangely." So do automatic dishwashers and Federal Express.

She liked the idea that you could track your package at every moment by satellite. All of them flowing into Memphis and back out again, as if blood into the aorta. There was something strange and perversely comforting in knowing where something was at every step from moment to moment. It was why people kept dogs.

She decided to get to work for awhile and warmed up a mug of water in the microwave and went to her desk to make client calls. White tea, her son called it. He thought it charming that she drank hot water all morning.

"In Russia it is quite common," she said. "Or so I have read."

"Jennifer drinks mint tea that she makes by soaking fresh leaves in cold tap water. The flavor is hardly there even though it takes her forever to steep."

It was near the end of their time together at breakfast. He was thinking of calling Jennifer to let her know the package had arrived but then realized it was three hours earlier there.

"So our lesson for today," she teased her son, "is that women drink water which they prepare somewhat strangely."

Then she paused and made a brief mistake, which thankfully seemed to go unnoticed. "Perhaps I *will* like this girl," she said, instantly wishing she had bitten off the "perhaps" before letting the statement go out into the air.

He heard the liking and not the perhaps. He would ever be partly a boy hearing what he wanted to.

In a sense he had turned it into a vocation. All day today he would occupy a conference room and listen with an ear trained for complication, disjunction, inconsistency, misrepresentation. He would listen for what he wanted to hear and when he heard it he would tally it. His craft and art was words treated as if they were mathematics. It doesn't add up, a lawyer would say, I'm afraid your lies multiply, we are divided in our opinions. The toothbrush stood as an actual troth against that mathematic. It was itself, a thing that was once where he was before and was here with him now. Yesterday and today closing into a single bulge like a paper fan.

She hoped her son had an eye for these things to match his algebraic ear. She sat at her desk and wondered what she could send Jennifer that would let her know what she knew of her. Some bauble in a white cardboard folder tracked by a satellite. A scrap of frayed satin from the binding of his favorite childhood blanket. No, something simpler, more here and there as one.

A matchbook, she thought, though no one smoked anymore and he hadn't any. But something like that. Directly thingly, haphazard until its selection.

Jennifer loved the smell of the library, of bound leather proceedings, pleadings, statutes, and judgments, although mostly she worked in electronic space when she researched, steering through the airless tracts of Lexis and Westlaw, tumbling down the steps of databases, fusing search strings into minor worlds like biotech halflings sewn together from scraps of DNA and alchemical poetry.

She liked when she could duck her clerk and research on her own. It was best when you knew the feel of law by hand, though she accounted leather volume and keyboard as equally palpable. What she didn't like was photocopy and fax. They swam through space sans fin or scale and thus were as bad as the abominations of Leviticus. Reading a fax was like reading a chalkboard after it had been erased.

There was no longer chalk of course. Everywhere in the firm there were whiteboards and dustless markers, some with a fruity smell and some smelling of furniture polish and kerosene. Or sometimes, when you were lucky, a conference room would be furnished with flip charts, grownup versions of kindergarten easels soon besmirched with the Kool-Aid smell of the fruity markers.

She loved it nonetheless, loved every bit of it. Loved it more now that she was in love with Malcolm. Law was love to her.

She didn't think this too philosophical, but then it wasn't really an idea she quite shared with anybody. Not even Malcolm, or not directly. You learned to subsume enthusiasms in your second year of law. You grew counsel, decorum, due deliberation, those words. And you spent hour on hour in law libraries.

It was alright. Counsel was a costume like a birthday dress. A girl learned early in life what it was to dress up, what it was to come home.

Or at least she had. It doesn't do to generalize, in law or girlhood.

She pressed the Enter key and submitted the search string. It went out into the unseen world like a new citizen, full of expectations and yearnings. Instantly the screen refreshed and already the hits began to clutter the space where it had cleared. It was raining wisdom in an ordered spatter.

It was a yearning that she sent him via FedEx. She woke to see their toothbrushes peering at her, their bristles looking out from the little slots above the rim of the cupholder below which their handles dangled like legs.

Two with one gone. It was a poignant arithmetic.

Malcolm, of course, had a separate toothbrush neatly encased in his folding Samsonite shaving kit ready always for travel. He was the soul of due deliberation. She, for her part, could not bear to part with a toothbrush once she broke it in and so swapped it back and forth whenever she traveled.

She sent this one to him as a forget-me-not. She could as well have sent him lingerie, women did that, or a perfumed hanky, as her mother might have. Her Italian grandmother had a rosary with a little window which was said to contain a relic of the true cross, a soggy little sliver from the beam which supposedly held Jesus at his crucifixion. She never liked to see it, it was frightening to think a thing could survive a being. Her grandmother had a set of tarnished and undistinguished spoons wrapped in newspaper from the old country.

That made more sense. The past was something. Her grand-mother's fingers were skinny, wrinkled, and olive and felt like velvet.

She suddenly wondered about his mother's fingers. His own were awkward, although my god you could never tell that to him. He paid attention to the particulars, had manicures with his haircuts, wore french cuffs, wanted to move up in life. Still his hands were boy's hands, even in his blue suits, especially on her body. Soft and happy, not graceless, not—my god—unadept. Maladroit perhaps. Athwart. Guileless and enthusiastic.

He was her youngest lover yet since she joined the firm after law school. For awhile she had wanted to be held by men who smelled of cologne and thought secret thoughts and were vain about their bellies. For awhile she wanted taxicabs and dress-up dinners and long showers and sweet-smelling soap while someone waited.

People said something had a life of its own. For awhile she had wanted to hold someone whose body had a life of its own, a history in it.

It was a phase and it ended. Malcolm was different. Courtly and self-important, he thought he invented love. It was charming but he needed finishing. She wasn't sure whether their relationship would last. Everything was just beginning. Love, she thought, works strangely.

What would his mother think of her?

She tried to imagine her in her apartment above the city, above taxicabs and treetops and people walking. There was a small terrace, she knew, and when Malcolm was a child he had watched an eclipse of the sun with his hand in his mother's, a little afraid and not quite comprehending anything but the strangeness of the mottled shadows on the terrace which he still remembered.

A Man on the Moon

In the course of calling every Ed Eyre in America, her father, Ed Eyre, conceived of a project for a book. It put them on the phones and email, Gwyn and her mother, thinking what to do if anything.

"What could you do?" David asked.

He called her Jane sometimes but this wasn't one of them. When he did so, she called him Tarzan. He wondered sometimes if she and her mother managed their love for him as they did their love for Ed. He thought they probably did, though with less history between them and, not to be in any way self-sorry, with less love as well. They had been married for only eight years. Gwyneth was thirty-eight, her father and her mother neared whatever anniversary it was, silver he supposed, for fifty.

"Well, that's the point exactly," Gwyn said, "But sometimes like anyone she has to talk that out. The hardest thing sometimes is to set your mind on doing nothing."

"The Hippocratic oath," he said.

She narrowed her eyes in the way she did when he didn't make sense.

"Do no harm," he explained.

He wasn't a doctor, as the commercial used to say, he merely played one on teevee.

"My god, David!" she would say whenever he would use that line, "You have to be ancient to remember that reference. You need to do some heavy time on the channel zapper like the rest of the men in America. Gotta update the commercial database, darling."

She also hated whenever he said, "Let's have a happy Fizzies party," although he had read in *Advertising Week* that the candy company was reviving both the brand and the campaign.

He wasn't an advertising man either. Ed Eyre had been and sent them a subscription every year as one of their Christmas presents. Ed was what you could call a character but never to his face. He was crazily handsome, always tanned though never from a booth, the kind of man who people were, literally, always coming up to ask if he was a movie star. Gwyneth inherited his beauty, she was, again, what you would call statuesque and a classic beauty. It made it tough on David and his mother-in-law.

Though there were compensations enough. Ed Eyre and his daughter were each compassionate and bright and, despite her intolerance for commercial jingles of the sixties, good humored and witty. They were the perfect family and that was the source of their dread, he thought.

Like every family they awaited a flaw, a fall, a fissure, the beginning of the end. He and Gwyn could apparently not have children. That seemed a start on it but they had worked out ways to deal with their situation midway between cooking up each other's genetic materials at a fertility clinic and adopting a baby through a classified ad with an 800 number in the back pages of *The Nation.* They would wait and try lovingly, regularly, happily, use certain herbs, certain positions, exercise, even pray.

What he was was a minister. Not a doctor or an advertising man, though these were clauses which, if spoken, Gwyneth would rush to fill with humorous objections, since, she would claim, a minister was midway between a doctor or an advertising man. David wouldn't disagree. It was true and also no one disagreed with Jane Eyre.

Sometimes she called him Reverend Tarzan, or Father T, though the latter of course made him think of the topknot and gold necklaces of Mr. T, the television star, and then naturally of the bloody-mary mix on airplanes, which was for some reason called Mr. and Mrs. T's.

"And that makes me Mrs. Tease," she said once, playing with the buttons on his shirt like a vamp in a movie.

It took his breath away. She was the wife he longed for whenever she was gone. They were happily married, words he never dared pronounce.

It was airplanes that first set her father on his project to telephone what he called all his homonymic doppelgangers—people who shared his name—throughout the United States. He had once met another Ed Eyre in an airport somewhere when the ground attendant paged the lounge and asked Ed Eyre to come to the podium and another man joined him.

They looked in no way alike. The other fellow was "crabbed and compact" Ed had said.

"Not crabby though, darling," his wife said.

"Like the tree," Gwyn assented, "Though that would suggest compact as well."

"Do you see how it is with them?" Ed has asked his son in law.

It turned out that the compact Ed was really quite a voluble fellow, smelling a little of whiskey, a laugher and an inventor. This was what Gwyneth's Ed, her father, wanted now to reach the other Ed about. He didn't have an invention, although he was very inventive, but rather for the first time in his life he had some real money to spare and he was looking to take some risks with an investment. He thought maybe Ed Eyre the inventor could point him to a project.

"You want to lose a lot of money? Buy lottery tickets," Gwyn said to him.

"Or give the money to David, for his church," her mother said, nodding toward her daughter's husband in the comically reverend way she sometimes did.

"She is never sure whether we are sexual," Gwyn once said.

"And you don't tell her?" he asked.

"Tell her what?" she said in deadpan, then laughed.

It was a fright for him.

"I'm not certain my church could offer even the promise of the sort of bonanza that Ed is interested in," David said carefully. The truth was he meant it. Let Ed do what he intended, the church was a pilgrim.

"You only have to ask, you know, David," Ed said sincerely.

"He can't ask," Gwyn said. "For David the church is mendicant."

"What's that mean?" her mother asked. Her name, unaccountably and aptly, was Grace.

Gwyn jumped right in, she never asked David to march out of swamps where she had brought them.

"It means that the church is a constant petitioner. A mendicant is a beggar," she said, "But one who waits patiently, both ever unworthy of and also ever grateful for and elated by what comes to it. It is like sex."

It was a fright to all of them. It was also, David believed, a holiness as well. It reassured him about the life she had to lead with him.

"It's what I'm looking for," Ed said, stopping them with the inadvertent comedy of it.

"Sex, daddy?" Gwyn asked.

"In a sense," David said, "Ed thinks of this money as profligate, on a sort of a fling. But he has hopes that it will return like the Prodigal Son."

It was as much a comedy. He wasn't the kind of minister who crafted tiny exemplums from scripture like someone carving wooden elves from a tree branch. There was a joke between them whenever he was acting too pompous and wanted to signal Gwyneth.

"I am reminded of scripture . . ." he would intone and she would laugh with him or at him as he wished.

"There you go!" Ed said, "Money on a fling exactly."

"I wish he would just give it to you kids," her mother said.

"There's plenty of that," Ed said, "We just have to keep from living too long. Anyway if the prodigal comes home, he can move in with Jane and Tarzan."

He and Ed Eyre the inventor had exchanged business cards but he would be damned if he could find it. Nor could he remember, when Gwyn and Grace challenged him, what airport it was, though it didn't matter a hell of a lot, as he said, since all the airports in the world were essentially the same.

"There's a bright tunnel and uncomfortable chairs, another narrower tunnel, a still more narrow tube, even more uncomfortable chairs, another set of tunnels, and then the sky outside a world away."

Ed could spin poetry sometimes out of flax.

"It sounds like how people describe after-death experiences," Grace said.

Gwyn studied David though he felt certain no one took note of it but him. He had struggled for years to understand what he believed was an out-of-body and maybe even an after-death experience years before following an attack of peritonitis. It was in graduate school and not at all why he had entered divinity school thereafter and so he was sensitive about telling anybody but Gwyneth.

"I was hit by a car once," Ed said. "I was just a little shit, in my twenties in the city. It knocked me on my ass on the sidewalk and then I woke up and walked away from the cop who was bent over me asking questions."

"My god," Gwyn said. "The things you find out if you start listening to your parents."

"Wait 'til you hear about his mistresses," Grace said mischievously, although nodding her head toward David like a genuflection.

Ed hated computers but when he saw an ad about nationwide phone directories on CD ROM he had a way clear to try to locate Ed Eyre, the inventor. He convinced Grace to show him how to use her 486 and then started going upstairs when she settled into Weather Channel after dinner. It was a good solution all around since his cynicism about her watching the weather sometimes got on her nerves and once he started making calls it kept them from being interrupted by incoming telemarketing calls after dinner.

He found Ed Eyre the inventor almost right away in Memphis. Maybe he remembered something in the back of his head, maybe it was luck.

"It was doubly lucky," Gwyn told David when she reported the events. "The other Ed Eyre miraculously couldn't think of an invention Dad could invest in. He had made millions since they met in the airport from some little switch that sits in a car engine. He was sitting pretty and he counseled Dad to put his mad money in municipals."

"I'll bet your mother called him first," David said.

Sometimes in bed she twined her legs around his thigh in the way that lilac bushes sometimes wind around the trunks of maples. He loved the powdery smell of her breasts. Amen.

Her father had decided to work his way through the electronic phone directories, talking to any Ed Eyre who would talk to him. It was after the conversation with Ed Eyre the inventor that he conceived of his project. He would ask each subsequent Ed Eyre he reached to fax him a memory of the first landing on the moon.

Gwyn and her mother worked through the practical aspects first.

Why fax?

He thought that more truly committed the other Ed's. When someone sent you something he was more apt to stand by his word. In any case faxes meant they could also send snapshots and other souvenirs and visible recollections.

Who would read such a book?

There would be a natural audience. It was a gimmick and a cross-section both. Sure people might treat it as a laugh at first, a novelty, but once they saw how an event could unite a scattered tribe of strangers sharing only a name, they would naturally begin to reflect on the meaning of history. It could be a runaway bestseller.

Why now at this time in his life?

Her father, of course, answered none of these questions directly, nor did his wife and daughter ask them directly. Rather they read the answers like someone reads perfumes or the evidence of a not-yet-cleared kitchen table. They looked for what he lived in.

There was an old joke between Gwyn and David. They were each fond of walking nights in the city or around the village at Ed's house on the shore and looking in others' windows. It was, David said, one of the few absolute differences between men and women.

"Women look to see how someone else lives," he said, "Men hope to see someone naked."

They had their share of seeing each over the years.

She had teased him by calling him an essentialist. He supposed that meant someone who believed that biological differences between men and women somehow led to other differences, whether mental, artistic, or spiritual.

"You left out sexual," she said. She was still teasing, he was relieved to hear in her voice. "Sex and gender are different," she said, the teasing turning vaguely flirtatious.

If Gwyneth and Grace had not interrogated Ed directly, neither had they conveyed their concerns directly to David. He didn't think it was self-congratulatory—or essentialist—to think that he nonetheless knew their concerns. He felt—this was dangerous, he did not know how to express it—that he understood certain women with something akin to what people called a sixth sense. It was a grace. Not his mother-in-law's name but creation speaking through being.

He thought they were concerned that Ed was worried about meaninglessness. *The Book of Eyres* was meant to be a testament; David knew this from his own interrogation.

Why the moon landing? What about Martin Luther King's dream for instance? Or the death of Kennedy?

Because the moon was universal, not that great men weren't. Because it was a source of joy and pride for all the human race that men had looked back down upon their earth and found the beauty. Because we owed the moon veneration to the extent that we had trespassed upon her. Because the moon landing predated all this damned technology and cyberspace.

Wouldn't there be people who couldn't remember or who hadn't been born yet?

People had memories of even what they never saw. These days photographs took the place of memory—photographs and documentaries. There was an eight year old named Ed Eyre in Wellman, Iowa, who remembered his grandfather telling how he stopped the car in the middle of a country lane and listened to the landing on the radio and cried. The boy said his grandpa told him that the moon wasn't out that night and so he couldn't see. He could only believe.

Could you really expect people to send faxes?

The boy in Iowa thought he could. He said there was a fax machine at his elementary school and the 7–11. Others would respond likewise. In direct mail a two percent reply rate was thought to be a landslide. He hadn't been involved in telemarketing and so he didn't know the success rates there, but he supposed with cold calls it was much the same for a good telemarketer.

Wouldn't there be more success as an Ed Eyre calling others?

"Not all Ed Eyres have that much in common," Ed said, "Just their name and their humanity. They all walk on two feet and lay down to sleep at night."

Sometimes when they slept Gwyn shifted her rump against him

with a grace and presence that signalled nothing less than the mystery of Incarnation. It made him feel very grateful and full of love and longing. Often when he stirred next to her she woke to him and embraced him and he was swallowed up in her spirit and ecstasy.

He woke to find her staring at the ceiling, her eyes moving as if playing out waking dreams. She had the sheet pulled up around her neck and the beauty of her face was resplendent even in her worry, her profile like a cameo softened by the light.

It was sunrise, the hour of Matins in the ancient church.

"I thought Matins was midnight," Gwyn said.

"The second hour, two A.M. originally," he said. "It moved to the first light in the middle ages."

"All humanity moved to the first light in the middle ages," she said.

She had Ed's gift of making silk from flax.

"Is it anything you want to talk about?" he asked her.

"Oh I want to talk," she said, "but there's nothing to say, really. I'm worried about Daddy."

"This Ed Eyre thing?"

"Oh that's just an anchor for it," she said, speaking the words into a nuzzling kiss at his shoulder. She breathed him in in successive draughts. It was delicious for him.

"He will die one day," she said. "This project signals that and it makes me sad."

David fought with every fiber of his being to keep from saying "we all will die." He fought successfully. He slipped his palm around her breast and cupped its weight like holding a sleeping dove.

"I'm sorry," she said. "I'm really a little out of sorts."

She kissed his forehead.

"It's fine," he said, "I just wanted to hold you."

As he began to withdraw his hand she held it against her briefly then let him go.

"I think mother's worried that he'll ask her to use her contacts for The Book of Eyres."

Grace had been an executive editor at a publishing house. She still went down to the city once a month and had lunch with friends in publishing.

"Did she say so?" David asked.

Gwyn made the nuzzling noise in the air. "No," she said. "It's just something she knows."

She kissed him again and rose from the bed, walking serenely pale and naked across their bedroom to the antique cherrywood bureau.

He wanted to call her back, wanted to tell her that he had not yet spoken his fears to her. Yet it was time to let her go, he knew. They had journeyed through this night and now she was into the quotidian, the work of the day, a fabric of care.

For awhile he lay in bed still terribly afraid. There was in his however fragile and uncertain calling the injunction to tell the good news of Jesus Christ. He did his best, he knew, and even knew that he did well in what he did. He often took consolation in the parallel construction of the prayer that asks God's blessing for what we have done and what we have failed to do. Yet his father-in-law's project stirred a great anxiety in his soul.

However comic and impractical and quirky and madcap, Ed had spent his life in reaching out to men and women. His advertisements, for good or ill, had bathed millions. Now in his retirement he ministered among a strange flock, making call after call where surely he was greeted as some sort of fool, the kind of man who attends a national convention of Smiths. Yet he worked his way across the silver mirror of the CD ROM with charity and patience and not a little self-mocking. He tended carefully to the faxes when they arrived, as indeed they had begun to, retyping stories of strangers and re-drawing illustrations with his own hand.

Gwyneth came back in to get something she had forgotten. He

could hear the shower running in the distance, sense the damp of steam already clinging to her like scraps of mist cling to the hills at morning. She wore as always a creamy silk kimono rather than a bathrobe. Gwyneth was the moon's name.

"What's wrong?" she asked and came toward him. "Oh darling, you are crying."

"I want us to have a child," he said. "I feel so useless, Gwyn. I'm so so sorry. I know it is all my fault."

He had no Jesus, no him or her beyond his failing light. Yet he remembered their child as if it were real, a vision not like a photograph but rather one of those ultrasound images of the babe in the womb that people often show you, ghostly evidence of possibility and immanence.

She held him and he sang her praises. Oh moon, he thought, oh dawn, oh love.

THIRD MEDITATION
Memory Picnic

I remember a long strike at the steel plant during which we pic-
nicked every day at the city park nearby us. My father was locked
into the plant, a supervisor by then, while outside union men
walked the picket line as he did once. Every day while he was locked
in my mother would take us all to the park, four boys, four girls, of
various ages, the food pulled in a wagon. I connect this memory with
another one in which my father warmed himself by a burn barrel fed
with wood from empty pallets. He was on the picket line then and it
was obviously winter not the steaming July of the picnics in Cazen-
ovia Park.

Yet I hold the two memories together in a packet. My father and
the burn barrel and the park where we cooked food over charcoal and
swam in pools that smelled of chlorine dumped from large steel barrels like the burn barrel.

People involved with computers like to talk about the synchronous and the asynchronous, delayed time and real time, as if they
somehow differed from one another. A chatroom where various people take on identities and genders, fantasies and ideologies, is thought
to be synchronous, its transcripts evidence of what happens in time
as accorded to time. A newsgroup or website where people post email
about a given topic, or at least under its aegis, is thought to be asynchronous, the time between composing and reading delayed.

The memory palace, the progenitor of the memory picnic, is a virtual space where events or ideas are stored asynchronously and retrieved synchronously. One imagines a set of dishes, for instance, in a kitchen, which is meant to stand for the 1976 Cincinnati Reds, a baseball team. On the plates are the names and perhaps even the statistics of given players: Joe Morgan, Ken Griffey, George Foster, players most baseball fans remember, Cesar Geronimo fewer of us recall. In order to recite the players of the Big Red Machine, you "walk" through the kitchen synchronously, examining each plate in turn, reading from its colorful contents.

It might be possible, I suppose, to populate the same objects with another layer of memory, say the kinds of meals my mother prepared at picnics over the years, ranging from the obvious hotdogs and hamburgers to goulash, fried potatoes and eggs over easy, succotash and pork chops, and more than once a whole roast turkey. It might further be possible to associate the foodstuffs with real events or long-dead family members, to mix and match, making the lima beans of the succotash into a list of irregular Latin verbs, and the corn a rosary of the names of chiefs of the Iroquois Confederation.

A harddrive works like this. A look over its surface with any sort of low-level disk editor will show you that its contents are interspersed and overwritten. High matters are scrambled into the omelet, mere registers share the names of the Caesars, both the emperors and salad variations alike. We have come to believe that our minds work this way also, neurons or something (we tend to be essentialist about this, thinking of the texture of cervelles or sweetbreads in brown butter) holding random memories, linked together by pathways or lightning storms or regions, perhaps the way a pomegranate holds its dark seeds.

It isn't necessary to think that these memories, which are after all always generated synchronously, are necessarily synchronous. It is probably not even necessary, as my ex-wife—a sort of psychic—often said, to think we remember only backward. The only time I remembered forward was the night before the Challenger tragedy.

I was driving down Franklin Street in Jackson, Michigan, just passing Under the Oaks, a micro park that was the site of the founding of the Republican Party (unless you are an adherent of the Ripon, Wisconsin, creation myth for that party). In any case it was evening, just after dark, and I was listening to a report about the upcoming launch of the Challenger.

It will crash I thought, my spine electric with fear. I should tell someone.

It was, of course, impossible to do so. Whom could I have called who would have believed me? Why would I have called anyone else in order to record this prediction, unless I believed, as I had no reason to before or since, that I was psychic? Why would I have any reason to believe this premonition, which then felt a certainty, was correct? What was left to me was the synchronous experience of watching as the shuttle separated and zagged off obscenely. That and this telling, which serves no purpose.

Seraph

S he could not have fallen from the plane by accident. The way
she soared and circled, soft as a seraph, lazy as a cloud, it could
not have been a mistake that pushed her from the hatch and
into the chalky azure of an early September sky.

I know. I was her only witness.

To be sure others saw her tumble, others flutter, others kick out
and sail, others her peregrine turns, others her drop, others her still
air dangling. But I alone saw her land, doughty as a space traveller,
precise, polite, and in some sense businesslike, if one can speak of
grace as a kind of business.

Among those who saw her tumble there were chiefly the authorities and travelers on the interstate outside town, the latter apparently setting off a magnetic field of cellular phone reports that a
skydiver was falling toward the high voltage lines that cross the sky
like a symphonic score beyond the power plant. These reports were,
of course, premature and in some sense spurious since the spot where
she fell and the one where she landed were miles apart and anyone
seeing her sail could know that she controlled her fate as well as any
of us, steering herself deftly, working the cords like her own mari-
onette. Yet people these days expect a horror, and, in any case, do not
the most of them understand gravity or the principles of flight? They
thought she would fall like a shot dove onto the power lines.

There was little to do. You can hide in the sky, at least in a small
town. She had come out at low altitude and opened her chute late,

this of course leading motorists to think she was merely falling. But the drop had the secondary effect of giving her refuge in a relatively small span of sky. While she could still be seen from some roads, it wasn't clear as her circles became concentrically smaller in exactly what neighborhood she would land. She looped into her own view of the world and mine and landed on tiptoe in her pretty and impressively engineered canvas shoes.

We are too small a town, of course, to have our own traffic copter and even the television station at the state capital, the one nearest us, leases its copter only during drive time and tragedies. The sheriff could have driven out to the airport and found an eager volunteer among the dozen or so flyers who spend all day at the coffee shop there, but by the time he drove out there and back she would have landed.

I cannot say how many others saw her tumble or flutter, but for me she came into sight as a billow and a rush. I loved how she hung there, feet slightly thrust out behind her, goggle-eyed, serene, the merest whoosh of wind flapping across the chute's sail.

Visitations from the sky are not unknown here. Like anywhere we have our share of the zealous who on long nights see aliens and dream of chromium beds and spotless impregnation. Somebody said that the sheriff's ex-girlfriend once saw three flying saucers hovering over the ridge near Deer Lake, though she and the sheriff parted ways not long after that. Another time a jetliner making the first leg of its hundred-mile descent into the big city airport upstate dropped part of a flap like shrapnel into a parking lot near an auto supply store. Fortunately, as they say, no one was hurt, although the state capital's news chopper did make a visit for that.

At the historical society in the old Grange there is a glass case with an exhibit about a comet that supposedly fell to earth back in the Indian days, although I suspect they mean a meteorite. The evidence for this occurrence is the journal of a long-gone local Civil War veteran who wrote down the recollection of an even older Indian acquaintance. The comet is also commemorated in a piece of beadwork of unlikely provenance, most likely the work of an earnest boyscout or the old codger, Emmet, who is our one-man historical society. Both the journal and the wampum belt are in the glass case together with a steely colored rock the size of a grapefruit that glints like blue glass in its hollows and looks like slag from a blast furnace. The historical society is open only on weekends and every other Thursday.

I know for a fact that my friend Herbert once had a hot-air balloon set down in his farm pond. He heard the thing blast and snort just over his house and looked out in time to watch it drift down over the old orchard on a beeline for the pond. By the time he got his boots on and walked back through the orchard to the pond the balloonists had moored the huge bright thing, tying it to a couple of the dead grey tree trunks that still stuck up in the center where years before he had flooded a meadow and woodlot to make the pond. When he came through the apple trees he saw the bare ass of a fat man sounding above the water and the sweet underwater ghost of a naked woman gliding away just below the surface of the pond. There was a third fellow, thin and ferretlike and fully dressed, in the basket of the balloon and he shouted a warning to the two skinnydippers. The naked woman swam back and scampered up a little rope ladder, slithering into the basket like a lovely fish. Then she and the pilot hauled the fat man in the basket and they cut the mooring ropes and drifted off in laughter and loud tongues of fire. Herbert says as long as he lives he will never forget the sight of her jackknifing into the basket or the sound of her receding laughter.

My seraph was fully dressed and unruffled, tucked in as neat as a hotel bed. She wore barley-colored coveralls of ripstop nylon with square zipper pockets on the calves and thighs. By definition this was what you'd call a jumpsuit, I guess. Her shoes looked like the kind of amplified sneaker you see advertised in the mall as rock climbing shoes or hiking boots. But since there is an athletic shoe for every sport from badminton to weight lifting, I suppose these could have been designed for skydiving. I only remember the touch of her toes and the slight grunt as her knees flexed with the impact and she straightened out and began to rein in the billow.

For a minute I had the unreasonable feeling that she had come for me or to me. It's something you can't really avoid feeling when someone slips down from the sky and lands on the lawn next to where you sit.

I know I watched her from well before she landed but I can't exactly say when I became aware of her or what I thought. I did hear the sirens start up long before she landed.

The canopy of her chute was sewn with alternating bands of orange and gold and brown, not a circle parachute but the top half of a tunnel like the canopies in front of apartment buildings in the city. The brown and orange bands made you think of monarch butterflies. It had been weeks since I marked the monarchs passing through. I always like them, their flights both distracted and deliberate, bound for the mesas of California and Mexico where they cluster so thickly they fill the trees like plump fruit and block out the sun like a Navaho umbrella. It is something I always wanted to see for myself, though we never got there.

It was why she jumped, I think, though I am sworn not to tell anyone.

"Why in the world did you do this?" I asked her as about the fifth in a series of dumb questions. I never really got a good answer.

My first dumb question was "Can I help you?"

She was still reining in the chute like a rodeo cowgirl and I had hardly gotten up yet from the lawn chair or made my way past the roses and across the lawn to the double lot next door where she landed.

The way I said it was like someone wanting to help you with your groceries at the supermarket. It made me feel dumb as hell the moment I said it but she didn't really pay it any mind. She played the lines in and stomped the billow down into a hexagon then dropped the lines in its center and began to fold the thing in on itself until it was a tricolored lump the shape of a loaf of bread or an old fashioned haystack.

I don't remember her unhooking herself from the harness. Maybe it happened so fast I didn't see it, or maybe I was still thinking she had come for me.

The sirens were getting closer now, exactly like in a movie, though you could tell from the prowling noise of them that they weren't really sure where they were going. I looked up and down the street but there were no neighbors out. The kids were back in school by then and summer was over and not many of us sat out in the sun.

"Where did your plane go?" was my second stupid question.

This one, thank god, she laughed at like anyone would.

"You know, I never thought of that," she said. "I always assume it goes back where it came but there's really no way I could know, is there?"

"No," I said, more dumbfounded than dumb.

By this time I had taken a look and seen how really dangerous it was. There were the normal lightpoles of an old neighborhood, triple strung now with telephone, cable, and power. There was the steeple of the Presbyterian Church, a pointy white thing without a cross. There were oak trees and larch and one huge and precarious poplar. There were houses up and down the blocks surrounded by stockade fences and pointy chainlink.

By now she had the goggles off and you could see that she was on the plain side of pretty, a little jowly but peach skinned nonetheless. I also noticed for the first time that she was winded—that too the definition of the term I guess—winded and looking a little scared.

I also noticed, but only after she slipped it off, that she had worn a soft helmet of the same barley-colored nylon. It was ribbed with padded tufts and looked like a modern-day version of the old football helmets or the headgear of the Russian cosmonauts, like Franky Frisch the Fordham Flash or Yuri Gagarin. Blond hair tumbled from the barley hat, somewhat artificially enhanced but spectacular nonetheless.

She was starting to pick up the chute. It had transformed itself from a monarch billow to a banded grub, very close to the shape and colors of a woolly caterpillar, the augur of winter.

"You could help by not saying anything if they ask you," she said.

I didn't ask her who they were. It could be anyone: sheriff, husband, doctor, the PTA, or the traffic copter.

The sirens were working their way closer through the gridded neighborhoods and toward the platted contour lines of our second-generation subdivision.

"Of course not," I said. "I have nothing to say, really, do I? I don't know anything."

"You could," she said.

"Say something or know something?" I asked. It was the third or fourth dumb question.

She had a wonderful laugh and she rocked back on her heels holding the bundled chute before her with two hands the way a pregnant lady sometimes seems to hold her belly up.

"Both or either, I suppose," she said.

"Why?" I asked.

It wasn't a dumb question, of course, though god knows, in the newspaper since that day, there have been dumb enough explanations. We live in a small town and the paper has little to report but superstition, prejudice, conventional wisdom, and speculation.

She did it to advertise her cosmetics distributorship. She was a feminist and she meant it to be an abortion protest. She was a Christian and she meant a prolife demonstration. She was a disgruntled housewife and she wanted to embarrass her estranged husband. She was part of a group of suburban thrillseekers who made a pact to do what each other dared. She was mentally disturbed. She was an untrained jumper. She fell out by mistake before the jump field.

I mean when you think it over there never was much question that they would find her. It was merely a matter of going out to the airport and asking who had jumped today. Still when she set off through the back yard next door and across the pipeline right-of-way behind the schoolyard and the old garden apartment complex, I really believed she could elude anyone if she wanted to. She hauled up her hump of sky and trudged away on high-tech feet, bottle blond and brilliant as the sirens truly neared.

Here's what I think happened. I think—I *know* as a matter of fact, since this much she told me in the moment we had before she went off—she had raised her family here and seen her children off and watched her marriage end for twenty-seven years all in one place and wanted to see it all for once from another perspective. From above. So on the day of her normal skydiving lessons she either jumped out ahead without telling her instructor or convinced him to let her, promising never to say he had.

That's what I think. Although I didn't ask, even when I had the chance, and didn't tell anybody what I thought had happened or even that I witnessed her landing.

"Why?" I had asked when she suggested I could either say something or know something.

She thought I was asking why I shouldn't say anything.

"I'd just prefer to have this for myself if I can," she said.

"You can by me," I said. "You've got my word on it."

"I'll email you and tell you what happens," she said and started off.

She was halfway back across the yard when I called after her.

"How will you know me?" I asked.

She stopped stock still and looked back at me with an expression so dear and so sad at once that it made my heart melt.

"I've lived twenty-seven years in this neighborhood," she said, "Raised my kids here, divorced, and stayed on. I see you walking mornings alone. I see you at dusk again. I know who you are. We are neighbors."

I said the dumbest thing then.

"My wife . . ." I said.

It was because she was so pretty and dear and I was so lonely, but there was no way to finish. There is nothing much to do but walk, walk and sit in the sun and wait for something to happen.

"Why in the world did you do this?" I asked her.

"Why do we do anything?" she said.

"Okay," I shouted after her. "Hurry."

For the first five years it was okay really. There was grief and getting used to things. But lately it has begun to hurt again. I am no fool. I know what is happening.

I don't have email but I'm thinking of it. My son and his wife are always nagging me and on television they show databases and newswires and multimedia movies of exotic places. There is nothing much to do but walk, walk and watch television, grow roses and read novels. The woolly caterpillars follow the monarchs. In autumn I put up chili sauce in Ball jars and a hot water bath like my wife did. In winter I snowblow down this whole side of the street. I am a good neighbor and I know how sad it can seem to be alone. In spring there are old tulips, half forgotten, followed by iris and lilies she planted long ago. Ants crawl over the tightly balled buds of the peonies.

Recursion, Virtuality, and Simulacrum

A mong the most difficult concepts for my students to understand (I myself do not understand them in a way that I can explain in logical sequence, but rather viscerally, in the way I attempt to understand and explain other difficult concepts, for instance how real choice differs from the comparable alternatives most of us think to choose among) are recursion, virtuality, and the simulacrum.

It is not merely a stylistic tic that causes a meditation on choice to turn up, appositively, parenthetically, in the sentence that lays out this difficulty. The three terms bring choice to the fore, require us to come to terms with how we choose to live and in what sense we understand our present-tense experience and embodiment.

Consider a continuing quarrel between lovers in which one party makes his way to have coffee with his lover, a meeting where they are going to try to work out what has brought them to this point. On the walk toward the cafe he mulls over the events that have led to this quarrel. She said it hurts her when he defines her. But she did not understand that it hurt him also when she claimed that was what he was doing. She couldn't understand that and kept pressing him to explain more. He went into a sulk as he always did as soon as she pressed him.

In this imperfect example he applies a certain retrospective understanding to each successive action of hers in his recounting. There are two systems of understanding moving in opposite directions, his responses to her chain backward in time, while both his representations of her actions and also his responses chain forward. The application of formulaic actions upon preceding terms is recursion, it generates a succession of responses in two different times ahead of the events in question: the envelope of time that follows the moment upon which the recursive formula is applied, and the present time that fills with the output of the recursion like the printouts spewing from the mainframes in cartoons of computer culture.

Even so the explanation got lost, itself at least momentarily buried in recursion. Yet the lover in our explanatory story must somehow function in such a world. Time may be moving forward and back, filling the gaps of the past and the slipping of the present into future, but soon he will arrive at the cafe and he will have to say or do something.

He begins to mull these options. If she says such&such, I will say this. But then she may say that I am only saying that because of so&so. It is like a chess game in which the pieces change their prescribed patterns, their motions rescinded, the pieces resettled, in light of following moves. I will tell her that when she responds that way it is like thus.

Yet he knows that as soon as he says that, she will say this other thing.

It is not exactly hopeless, they will have coffee and they will say something, but it *is* virtual.

Virtuality has slipped from something we could understand (a dictionary I consulted uses the example "the virtual extinction of the buffalo" as an example, they were so far gone they were as good—or as bad—as done for) to something we think to understand as adjectival. The virtual is pretend, virtual reality is a picture of reality. It is something that goes on in your head, the second of the old senses of that word which we once thought we understood.

Contemporary virtual reality likewise goes on in your head, or so we think (it certainly, at least currently, goes *on* your head, often the VR participant wears a helmet like a deep-sea diver, or immense goggles and gloves). In virtual life as we have come to understand it, the images—blinkered and blinking—form a depth, sounds surround and enhance it, and we endow some input device with a feeling of movement (as if following an arrow in flight) or we actually wander about the laboratory or amusement arcade on the tether of the machine.

Or we walk toward a meeting with a lover over coffee. We imagine that what we imagine will have happened has really happened, is happening.

Our friend arrives at the cafe angry. This will not do, he cannot have his emotions so defined, he cannot live in a relationship in which he is not as free as his lover is emotionally. It isn't fair that everything he says or does is thought to conform to some template while she alone escapes into a life of authenticity.

These experiences, the woman whom he describes and challenges, are simulacra, copies for which no original exists. He has stirred up a series of propositions and events recursively, creating a virtual her whom he addresses as if she were the she whom he imagines her to be. The woman at the table may seem the twin succubus of the simulacra but she is someone else entirely. She bears no resemblance to the imagined other. In fact she is not there at all any longer.

He looks about the tables wondering why she has not shown up as yet for their appointment. He wonders if he has remembered the cafe correctly, if he has somehow mistaken the time.

She has disappeared before his very eyes. She has never been so present to him.

What he will do under these circumstances is of paramount importance to him and us. If he can find a way to act as who he really is, and engage her likewise, there is some hope for him and us. Fiction exists in this space, the potential for authentic action and identity. It is neither recursive, nor virtual, not a simulacrum. She waits for him afterall. Do you see her?

Another Land

She had dreamed of the same land three days in a row, her mother disclosed late in the phone call, offering it as a sort of curiosity. It was not much different from the way, when her daughter was younger, she would isolate unusual buttons among the scattered pool before them. Look at this, she would say, her long finger skating out on a pearl disk from the pedestrian throng of buttons, or an obsidian square with a gold wire eyelet, or an enamel button with a blue goose standing upright at its center and done up in a bonnet. There were always things to see, the button jar like a collection of dreams, the sky full of sleeping forms by day and silver stitches at night. Now they were each older and sometimes it seemed there was less to see.

"Everyone has recurring dreams, mom," her daughter said.

"These don't recur, they connect," she said. "Anyway you learned that from me."

"I learned everything from you," she said and they laughed and that was that for then.

Three weeks later her mother called to say that the land from the dreams had been in the travel section of her Sunday paper.

"That makes sense," her daughter said. "These things run in cycles. Countries have publicists like movies do. You probably saw an article in some other place before your dreams; now there's another one in the paper. It's like when you see the same movie stars on every talk show and in all the magazines."

Her mother wasn't all that old. She had her wits, in fact she was quite sophisticated, a bit more hip in some sense, she thought, than her wonderfully earnest and accomplished daughter.

"How wise you are," she said.

"Oh mom, I'm sorry," her daughter said. "So what country was it? Bolivia?"

"Cathay," her mother said.

"Cathay is—was—China," her daughter said.

She made it sound like checkmate.

"You make it sound like checkmate," her mother said. She was used to speaking her mind and they were fond of each other, real friends by now.

"Well, I mean," her daughter said. "I got off on the wrong foot and now everything I say is impossibly colored."

She liked that phrase. Impossibly colored. It suggested curiosities to her: lime orange, red bear, yellow housewife, brown green, ruby tree.

"The paper *called* it Cathay," her mother said. "Maybe it has become the name of a province or something."

"Like streets in a subdivision?" her daughter suggested. "Cherry Lane and Cypress Circle."

"Like stage names," her mother counted. "Meryl Streep or Lola Montez."

"Meryl Streep is her real name. She was a Vassar girl. And who on earth was Lola Montez?"

"Who in heaven," her mother corrected her. "Max Ophuls made a film about her. It was an obsession."

They often talked like this, like two women doing a jigsaw puzzle together, putting pieces here and there in different sections, sometimes working along the same edge together, other times working independently but in tandem through a patch of furled leaves, the face of a spaniel, or a Van Gogh sky.

"There is an airline with that name," her mother added.

"Lola Montez?" her daughter asked.

They laughed.

"It flies to China," her daughter said, "Cathay Air."

"Sea birds fly to China as well, and dolphins," her mother said. "This Cathay was the identical world I dreamed for three days. I recognized avenues in the photos."

"How can I doubt your dreams, mom?" her daughter said. "You dreamed what you dreamed."

They had other things to do then and so they let it rest. They loved each other and talked at least twice a week.

No dream or every one is unusual, of course, on the first night. Yet in recollection this one did seem unusually colored. Most dreams, even of expanse, seemed to her cramped by the sensation of the mind itself working away underneath. It was as if you were riding in a plush sedan, sunk in a leather seat above balloon tires, your finger tracing the faux wood grain, and yet there was always the sense of the engine, an ingot of purpose churning away before you in a metal casket suspended above the road. Your dream mind blinkered and, almost imperceptible, narrowed your vision of any dreamt ocean.

This dream, to the contrary, was wide and light filled and different. Parklike and sun bathed and simple were other words that came to mind and which she shared with her daughter. The dream was not at all unpeopled—there were people she would remember all her life, a woman wheeling an infant in a forest-green metal stroller the color of Italian buses for instance; or a busy young man with an earnest, pleading face, a hook nose and dark black hair the color of a plum—but it was unhurried and undemanding. There were avenues and she made her way along them. It was not at all as one makes her way in dreams, shadowed over by distant purpose, the sleeper floating over it all like unseen cumulus. No, she moved through these streets intentionally, conscious and yet lighthearted, as if on holiday or in a dream. She had felt this way often enough in waking life. She remembered once, for instance, crossing a broad avenue beneath towering trees in sunshine on a spring afternoon in Saint Louis on her way to a tea for Adlai Stevenson in the sixties. It was a feeling of complex thereness.

She was making her way to a mansion, a colonial structure with broad lawns and cream-colored colonnades she knew. Hers was the knowing of someone touring who sets out in midmorning after a breakfast of reading guidebooks, the half-certainty of the recollected reading increasingly combined with a growing recognition of the actual place as it unfolds. It was not at all the haunted, echoey knowing of dreams. Instead she felt rather good about herself, somewhat uncertain of her path but confident that she could find her way out of any detour or misstep.

And indeed she did, emerging by a back way at the place where there was a lozenge shaped garden pool, very Italianate and lush, its geometric surface still as painted air and exactly the color of lapis lazuli. It was this place she recognized instantly in the travel section photograph, although the newspaper photo was taken from the main approach on the other side from the way she had come and the garden pool was thus at some distance. Still the lozenge of lapis lazuli was unmistakable, like seeing an old ring your mother wore somewhere.

Her daughter listened to the description patiently enough. It had been months now since the initial call where her mother told of the dreams.

"Six weeks," her mother said.

"You see what I mean?" her daughter asked. "You are obsessing. How in the world could you keep track of the number of weeks since you had certain dreams?"

Six weeks and three days since the first dream. Things fell in threes.

"It was vivid," she said, "Lapis lazuli. Bright as enamel and quite placid."

"I believe you," her daughter said, "I do. But what are the chances that you actually visited the place once on one of your tours and don't quite remember that now when you see it in the travel section. Do you know what I mean?"

She wanted to know if she was losing her mind.

"What would it mean if I could answer yes?" her mother asked. "If I can say I am losing my mind, it's like the old riddle of the two Greek villages where one village always lies and the other tells the truth. You never know."

It was a terrible confusion. She had left out the part about what she was thinking—how she thought that her daughter was asking her if she was losing her mind. As a result her daughter thought this an answer to the question of whether she might have visited Cathay and then forgotten, as if someone could. She didn't try to explain but rather named their common worry.

"Alzheimer's," she said instead. "It means someone else's house in German."

There was always this between them. The fear of the inevitable slip. Either down the stairs and a broken hip or tumbling through memory into senility. One aspect of their friendship was that, after years where they came to be able to talk about distance and the death of her husband and lovers and the birth of generations of their respective babies, she could now talk to her daughter frankly about aging and lovers and distance and the inevitability of loss and death.

"I swear next time I'm going to call that woman psychologist on the radio," her mother said. "You know, the one with the beautiful name, goddamn it." The accusation of slipping memory broiled her. "Susan Forward!" she said.

"She'll keep you on hold for hours. She'll tell you it's normal. I almost always answer the phone. I acknowledge the unusual. I'm merely a skeptic."

Her daughter was afraid her mother might book a flight to whatever Cathay it was that the newspaper featured. It wasn't, she realized suddenly, that she didn't want her mother to travel, she had been a great traveler, but rather that she didn't want her to risk the temporary certainty of her dreams against mere reality.

"Don't worry. I won't do anything foolish," her mother said, "I was actually thinking of a CD-ROM. There is no need to fly there, I assure you. I have been there in my dreams. I'd rather fly to Argentina."

It was another confusion, she mistook CD-ROMs for virtual reality. Her mother was a great fan of the Discovery Channel and so she had imagined she could somehow walk through the space of Cathay and revisit her dream within a computer. The variety of her mother's appetite for adventure always caught her daughter by surprise. Years before it had been a tale of entertaining three lovers, all named Ramon, in separate rooms of a grand apartment on the Costa del Sol. Not sex but conversation and choreography: cervesa in Salon A, vodka martini in the parlour, tea with lemon on the terrace. Not a Feydeau farce but As You Like It.

"Surely they knew," her daughter had said.

"Of course," her mother answered. "But they were gentlemen. The girl who worked for me looked in on them while I was elsewhere. She was pretty to look at and she brought them bowls of olives and salted almonds. I visited each room from time to time and laughed at their jokes. I, too, was very beautiful in those days. Each of them felt loved and welcomed. It was an effort to juggle and nothing I'd anticipated or want to do again, but we all survived." She paused dramatically. "They left as they had arrived, at separate times."

There had been a naughtiness in the pause that let her daughter know that one Ramon left much later than the others. This was during the time when her mother thought her daughter's life needed romance and possibility instead of diapers and a garden with lavender bushes tickling her knees. They had been friends for a long time. Now it was CD-ROMs.

The dreams were more linked than a continuation of some story. The second, of course, at first had seemed a dream of a dream and she almost woke from it on that account until she realized that the locale wasn't some signal of resumption but rather her real vicinity. She had almost stopped a rather imposing matron for directions, a woman in a black silk business suit carrying a coach bag, but she realized that she had no particular destination in mind. It was a relief to recognize her freedom but it was soon followed by a mildly annoying sense that she had forgotten something she really wished to see.

Once again this wasn't the same thing as how you think you've forgotten something in a dream. There was no sense of mediation, no sense of seeing yourself. The dream and her real life were interwoven and equally available, even neighboring. In fact she had briefly considered going back, either to the hotel in Cathay or her own bedroom, to look for what it was she had forgotten. Although whether this happened in her dream or in retrospection she could not say for sure.

"So let me try to get this straight," her daughter said. "You think that your dream world and the newspaper Cathay are places you can go to by different means."

"Not my dream world but the world I went to in my dreams."

"It could be," her daughter said. She meant it.

Who could really know. Her mother knew mysteries. They each did. She wanted to hear about the third night.

"You haven't heard about the second yet. It's not a story, not at all that sense of second and third. It would be as if you said 'Now tell me Cairo.' One doesn't follow another, except of course that I walked from place to place."

"Well, if they aren't parts of a story," her daughter asked, "then why insist that I hear about the second before the third?"

It was a wonderful question, an important question. Yet it left her mother feeling muddled and a little afraid. Perhaps her insistence on the reality of the place was a delusion. Perhaps in her seemingly clear consciousness of the dreams there was some subtle story, buried unrecognized and yet insisting on itself. She found herself filled with simple but unreasonable longing for a way to revisit her Cathay. It was how sometimes you momentarily felt an urge to telephone someone long since dead.

"I wish CD ROM were what I had thought it was," she said.

"You mean VR," her daughter said. "It's better to dream."

The woman in the black silk suit continued on ahead along the block next to the palace grounds but then immediately crossed and entered through a wrought iron gate into a small park neighboring a shady street of apartments. It would have been good to follow her but she remembered that the place she wanted to go was several blocks up ahead. The morning traffic noises were a muffled music and the air smelled like sweet hay. If you listened closely there were locusts humming under the police whistles and the screech of bluejays.

In the third dream she was fairly weary, which seemed to signal that she had been walking for some time and in fact that the three nights dreaming perhaps comprised one journey. Though not, as she had told her daughter, a story.

"It's unreasonable, I know," her mother said, "but I wish you could go there with me. To Cathay."

Her daughter didn't ask whether she meant the Cathay of dreams or the newspaper. Nor did she say that she wished the same. It was good. They knew each other well.

"I decided I wanted to find apples," her mother said. "Do you know the way you sometimes want nothing other than a perfectly ripe apple of an unusual variety?"

"Apples always bore me," her daughter said, "My memory of them is like wood. I get them for the kids and sometimes it's a surprise when you bite into one. I like the names though: Gala, Winesap, Rome and so on."

She was in a suburb and she wanted apples and so looked for a greengrocer, although she had a strong sense they were out of season. Again, she insisted, it wasn't a dream sense but rather practical knowledge of the kind you have when you are confronted with supermarket barrels of shiny apples in August. It isn't in season, it isn't in rhythm. It wasn't a dream.

Although there was in her mind no certainty about the seasons in Cathay. It was possible—a child's knowledge of a tunnel through the world—that everything was topsy turvy. Drains ran counterclockwise, apples were ancient in different months than in the world above.

You could look up and see the blue of the world above, as if looking through a mailing cylinder.

For a moment she was afraid that all this meant she was dying. They each were. The sentence equally applied to each of the women, itself a tunnel, a cardboard cylinder. Each thought: For a moment she was afraid that all this meant she was dying.

If that's so, her daughter thought, I'll leave my husband and my children and I will run off to Cathay with my mother. We will search for apples and sit in a park.

If that's so, her mother thought, I will have an answer about whether dreams and the real world match at certain edges. I'll follow the woman in the black silk suit and we will have a conversation over glasses of sweet wine. We'll eat apples and salted almonds; we will laugh and love each other.

Instead, of course, they talked some more about this and that. In time as usual the talk itself took on the quality of a dream or the memory of a long-ago vacation.

"You shouldn't let me keep you with these stories," her mother said. "Maybe I'll write out a map for you to see. In any case I'll send you the pages from the travel section as soon as I can find a color photocopier."

"That would be good," her daughter said, "Although I suppose I could write to the *Journal* and ask if they have back issues available."

They promised, as always, to call each other soon. They each looked for fruit as soon as they were off the phone.

FIFTH MEDITATION
Amusement Parks

From my youth I have had a recurring dream of death in which I sight the lights of an amusement park along a shore at night while overhead the sky is full of flying machines of all sorts— zeppelins, biplanes, jets, manned kites, and the drooping silver balloons of high flying explorers, the shape of spermatozoa or teardrops. In the dream there is a growing sense of delight and expectation, a carnival feeling precisely. Suddenly I know I am dying and still there are the chase lights, the calliope, all the flying ships.

The internet sometimes seems a momento mori as well, it too an amusement park appearing in the flow of things.

There were two ways to reach the amusement park at Crystal Beach, the first by car across the Peace Bridge and along Ontario Highway 2 and the second and preferred route over Lake Erie from Buffalo to Canada on the Crystal Beach Boat. Woman or boy, man or girl, always we called it by the full name, so: the Crystal Beach Boat.

Below decks there was a snackbar which, for economic reasons, we were never allowed to patronize. The main attraction there was a view of the engine room and the huge brass piston turning on a thick black arm. It was the Crystal Beach Boat that came to mind when I first read Henry Adams's account of the virgin and the dynamo. We would watch the huge arm and the piston for moments on end but then rush up the iron staircase to the deck when the park came into focus in the glare of sunshine above the horizon.

What is the internet? At some level it is a series of documents, the most of which (a number of them are generated "on the fly," spawned from databases and given forms by design algorithms) could be printed out in sheaves and baled in a warehouse.

I once saw a Senate hearing on CSPAN in which attorneys for a drug cartel explained the enormous logistical difficulties they have transporting and storing truckloads of money. Rats and other vermin inhabit the warehouses, mildew wrecks havoc. There are safe houses along the border in which every room but the vestibule (in case the authorities should ring) is stacked floor to ceiling with money. It is very hard to turn this money to anything useful: you fence it in bales to mega-corporations who turn their eyes from the obvious. How much money do you have on your person at any one moment? a Senator asked. The witness was briefly confused. You mean on my body or with me? Either, the Senator waved affably. In my pockets I have about ten thousand dollars, in my briefcase just short of a million, in the trunk of my Mercedes twenty million or so.

It could just as well be water. Coca powder flowing along the jetstreams and the actual waters of the equator and inland waterways, dollars likewise returning. Their papery thinness without the papery beauty of poppies.

What flows. Time and the river is a constant answer of literature. Water as well as light. The image of the thanotologists is a tunnel of bright light beyond which the dead await to greet you. A web page, for instance, the death of an instant. "I see in you the estuary that enlarges and spreads itself grandly as it pours in the great sea," Walt Whitman wrote in "To Old Age." It is as this broadening edge of flow where Samuel Beckett likewise eddies in "An Abandoned Work": "Just under the surface I shall be," he writes, "all together at first, then separate and drift, through all the earth and perhaps in the end through a cliff into the sea, something of me."

Anthology, by now everyone knows, is from the Greek for a gathering of flowers. *Antho* or *anthro thanatos, agrostemma,* the rose of heaven. The web contributes to our forgetting, the flowers strung together mean to instill recollection: Forget-me-Not, Morning Glory, *Amaranthus caudatus* (Love Lies Bleeding).

Saint Someone

As they were having dinner his wife asked him whether he had ever known a saint. The question came at one of those periodic still points. They were seated and the ceremony of menus was over and they had nothing to say. This happened occasionally and usually they lived through it together, waiting until one or the other of them recalled or was struck with something or until someone attractive to either or both of them came under notice. The mild fantasy of watching someone together from a distance awakened an eroticism which seemed the engine of even their most commonplace conversations. Food itself, or wine, often had the same affect. Memory likewise. The silence would pass and they would talk together in the ways they were used to.

Did you see her legs. There is a wonderful lemony undertaste to this Sancerre. Where was it we saw the blue lighthouse. The arugula is bitter as an autumn kiss.

They were not comic figures. They ate out only as often as most people and often talked of simple things, what went on or who had called on the telephone. Often they would laugh at their own indulgences or pretensions, the lemony undertaste for instance. Equally often they would delight in the unexpected inspiration of a phrase like an autumn kiss or the memory of a lighthouse. Generally they were happy.

On occasion, however, a still point between them would threaten to iris out into a void. There would be an abrupt fear that life had turned suddenly and irrevocably stale in the way a green leaf sometimes desiccates and crumples ahead of season. Perhaps they had outlived their love for each other. Perhaps there had never been enough to sustain them.

The food would seem tasteless and ordinary. People around them were awkward and unattractive. Wine tasted of must. They would find it difficult to meet each other's eyes. Silverware scraped annoyingly against the china; the sound of the other's chewing galled them.

This had not happened. The still point had lasted for an interval and then she asked had he known a saint. He did not answer the question for nearly two years and by then, of course, he had to remind her of the context. Meanwhile in the previous time, on the original occasion of the question, they had found a breeze to stir the stillness.

He could not, however, recall what they had talked about.

"My god," she said, "You don't expect me to, do you? I can only barely recall the saint stuff, let alone what we talked about after you had no answer."

"I had an answer even then," he said, "I just didn't have a name."

The waiter brought wine, Rabbit Creek Barbera, spice and rubies. They were out for dinner again, nearly two years later. It would be a good story, he thought, if they had returned to the same place and he had finally answered her question there.

"And so . . ." she said, "Our saint for the day is . . ."

It was very witty the way she said this. They were having a lovely time, talking easily, enjoying each other.

"I still don't know her name."

"*Her* is it?" his wife said. "No wonder you couldn't remember."

She was teasing him but it pained him. He had found the woman very attractive once. She was the friend of his lover during the days of Kent State when Washington Square thronged with riotous crowds and street singers and huge papier mache puppets of General

Westmoreland and Nixon. They all worked at the university research center together: he, his lover, the woman who was a saint, a great many others.

They were best friends, the saint and Linda, his lover. He could almost recall her face and thought he could recall aspects of her body, the curve of a certain skirt, the crisp cotton of a blouse and the lace pattern of the brassiere around her full breasts. Khaki pants. A pearl tweed wool skirt in winter. Her laughter, she and Linda laughing in the hallway outside their cubicles. It had pained him that he could not remember her name, truly pained him. It was something smart, quite contemporary then, Julie or Kim, or Karen, but these names were not her name.

"Because you loved someone that does not make her a saint of course," his wife said, "although loving you would make anyone a saint." It was really a very clever remark. She was teasing still, although more seriously. She wished to know what had prompted this insistence, persistence really, over the course not just of the two years since her question but, really, a life.

She had not come into his life until the middle of each of theirs. Like everyone, they each had married others. Like very few they had found each other. Yet there was so much taken for granted, so much yet to show itself. They married each other like water and at some depth there were smooth stones, time worn, burnished, lasting.

"I think that she died," he said. "I have a very strong sense of her death. It is as if I had seen her death notice almost by accident while reading the newspaper and once that happened it seemed so unlikely, so utterly impossible, that I could not keep it in memory."

His wife knew as well as anyone ever had when he was truly troubled. No matter how foolish it might seem to others, or even to herself, she knew and valued it, sometimes even despite the unlikeliness of what he was feeling. He loved her for this.

"Jesus," she said.

He sipped the wine and let its fragrance fill him.

"I mean it might even be the case that I attended a memorial service for her and then put it out of my memory," he said. "A wake or something. It's that strong a feeling."

"Deja vu," his wife said.

"Not really," he said, "I don't think it is deja vu when you remember something you aren't certain happened."

He imagined the unnamed saint in a linen dress within a pewter coffin lined in blue satin. A young enough woman that it seemed tragic. There were a few others there from the old days at the research institute who had also seen the notice, though not Linda, not his old lover. She had, he imagined someone telling him, moved very far away, somewhere exotic. Northern Africa. Morocco or somewhere.

None of this happened. It was a flight of fancy, an inverted sort of melancholy, the loss of his past. He wished he could remember the woman's name.

"We were not lovers," he told his wife. "She was the friend of a lover. I think I've told you about her, I know I have. Linda. In New York during the seventies at the research institute."

"The one you thought became a cartoonist," his wife said.

"Yes," he said, "She is in Morocco instead. Or at least I think so."

"I thought you told me she had moved to Bermuda."

It was quite a different place but he supposed she might be right. He could have said this once.

It was strange to enter this conversation after nearly two years. For him, he was certain, there remained an immediacy. It was as if he had carried a small stone in his pocket for years, not quite recalling the beach where he collected it, yet knowing the feel of the icy water, the smell of salt and kelp and distant pines. His wife would have less of this feeling. For her the original question had been a conversational gambit, a gesture toward him and against the still point. For her it would have been lost in the chaff of history. The immediacy of his memory of the conversation must have been disorienting, much in the way people describe the experience of receiving a letter from a dead man, something posted before his passing.

He tried to explain all this. She for her part had a vague sense that he had not only told her a story of how a beautiful woman from the research institute had once been frightfully kind to him, but also that he had at one point told her the woman's name, mentioning her in conjunction with his lover, Linda.

She drank the ruby wine and tried to recall exactly. The wine left a not unpleasant tannic taste, numbing the back of her tongue. She longed to kiss him.

"It was a very strange thing," he said, "although all things were strange in those days. The streets boiled with protests. We were wild with love."

She had been Picasso's son's lover, the wife remembered. Not she herself but Linda, his lover in the days when the streets boiled. He had won her away from Picasso's son though there had really been no contest, no struggle, no triumph. The woman Linda simply chose him rather than the Picasso.

Like that, she remembered. In the original time, during the dinner when she first asked the question about saints, their talk had somehow turned to a lecture they had attended at the Museum of Modern Art. It was a lecture about the future and technology, and a Dutch artist—she was fairly certain he was Dutch—had shown a movie of a man riding a bicycle through a city made entirely of words. It was the first she had heard of virtual reality. They had talked about that.

She considered whether to tell her husband this news but because he was already primed to tell the story of the lost saint she resolved to keep silent. It was not a movie, she knew, but a video the Dutch artist had showed, a video of his work, what they called an "installation."

"One time," he said, "Linda and I went to her friend's apartment—I hate it that I can't remember her name!—It was in one of those neighborhoods just above the village. Not quite Chelsea or Gramercy, someplace in the near twenties or high teens, none of this was chic as yet."

Despite his meander she knew his story would proceed in a single line, end to end, and the prospect of this threatened to bring about its own version of the still point. She feared falling into a quotidian abyss, a relentless march of details linked only to themselves but to no meaning. It was both fascinating and horrifying to anticipate this abyss and yet to be unable to stop him, unable to make him linger in the details of a lower Manhattan streetlamp or the fingernail of the martyress (which was how she now found herself thinking of the lost saint, martyress seeming a word filled with softness and loss like violet or orchid).

They were not lovers, he and the saint, although there was a sense that they might have been. That fact offered a preparatory aspect to the anecdote of her sainthood. The anecdote proceeded with an understanding that she had been kind to him at a time when he was helpless and under circumstances where she might as well have begrudged him his lack of performance or his presumption depending on whether or not he had misread her intentions.

"Linda and I went over there and then Linda had to leave."

"Kips Bay," his wife said.

"Oh no," he said. "That's well uptown, I'm sure, and east, Kips Bay is on the east side."

"I'm sorry," his wife said.

"About Kips Bay?" he asked. It was banter. It was how they sometimes joked when they had dinner together. He would tease her by claiming to misunderstand something she had said.

"No, about interrupting," she said seriously.

He had misread her intention. She hadn't meant to joke. He went on with his story.

In the story Linda left to do something and she was going to come back after some time. Hours. It wasn't the kind of situation where you might not be certain when she would return. She would be gone for hours—perhaps she had to work late, although that didn't make sense insofar as they all worked together at the research institute. In any case there was a clear sense that had they been so inclined, her lover and her friend, there was time to accomplish this without fear of Linda discovering their dalliance.

Perhaps she had even made a joke about this, they were always cracking rude jokes. Something like: If you two ever considered getting it on together, now's your time.

In those days it was well before the time that you could not be sexual in what you said. Still he imagined that her friend, the saint (what was her name, damn it?), might have felt uncomfortable. After all it was her apartment and her friend and her fellow workmate.

Not just her workmate, he was actually her boss in some sense. In the sense that he was the office manager at the research institute and she and Linda were each research assistants and so under his sway, as it were, his jurisdiction so to speak, although that was much too strong a word. Actually he had little to do with them. They cared for themselves, managed themselves. They would linger in his office and crack rude jokes, lunch together, flirt. A flirtation led to he and Linda becoming lovers, sadly enough for Picasso's son.

They smoked dope, he and the saintly woman. "We smoked dope," he said. "Once Linda left, her friend—our friend, this saint—asked if I wanted to smoke and we smoked dope together, side by side on her sofa. A folding bed, really. Like many people, she had a studio apartment and so the same room was a bedroom and a living room."

"We lived in a studio apartment once," his wife said.

"Yes, of course," he said.

He knew as well. It was how you told a story sometimes, filling in details.

The danger and the excitement had gotten to him. In the story that is, although the truth was that he felt danger and excitement in telling it now at dinner, two years after the question, nearly two decades after the event. He felt a thrall and a rush at once: those were the words for it.

Perhaps it was the danger and excitement or perhaps it was sitting alone together on the sofa bed in the studio apartment, or maybe it was simply bad dope—or good, sometimes unexpectedly good dope could have the same effect—but in any case he freaked. It had a bad effect. He was filled with panic and dread. He sweated and yet fell into a terrible chill. He could see her violet eyes above the joint as she sucked in the smoke. She was one of those people who held a joint underhanded, cupping it in her palm and taking in the toke slowly, as if breathing in through an oxygen mask. The way her palm curved reminded him of how a lover would curve her hand around your penis. She had lovely breasts. Her eyes squinted from the smoke and she smiled through the haze. She held the joint out toward him. He was in a panic and she did not know this. He stood up abruptly and the room spun severely. He sat back down and shook with the chill. Then she knew.

"How awful," his wife said. "How vulnerable you must have felt."

"She was a saint," he said. "She knew instantly and she commenced to address me in the most gentle way possible. She talked me through the hours until Linda came back, both acknowledging my fears and at the same time insisting that I understand I was capable of functioning. It was very canny, very caring."

It would be impossible to tell the story to his wife he knew. They loved each other like water but there were formless things beneath any surface, dark forms only dimly perceptible above the bottom.

More importantly he knew the story as he would have to tell it would sound foolish and comic. It *was* comic, just a routine event in the lives of those years, a man and woman smoking dope and one of them freaks. Sexual tension. Kindness.

"She never gave in to my fear," he said, "and yet she never discounted it. We left the apartment and went down outside for air. It was very difficult for me to walk down the spiraling stairs. She was the measure of calm. 'Let's walk two steps and you can rest,' she would say. I'd cry about how foolish I was to have this fear. 'Not foolish at all,' she would say. 'It's awful when it happens. I feel sorry for you. I'm sorry it happened.'"

She would caress his forehead, hold him to her breast, standing just below him on the stairs so that where he sat his head was just below hers where she could cradle him.

"What are you thinking?" his wife asked.

"How she held me on the stairs," he said. "It was more loving than it would have been had we had an affair while Linda was gone."

She would let him sit awhile but then insist, gently but insistent nonetheless, that he try again to move, that he try to regain himself. It was this that seemed saintly to him, how she both allowed his experiences and insisted on the possibility of transcending them.

His wife was like this. Even now as she sat and drank the wine and waited for him to continue.

She for her part felt a tinge of jealousy, not so much about the encounter between her husband and the young woman in his past, but rather that memory could pool in such a fashion, giving you glimpses but then covering over the smooth stones which lay in its depths.

It was clear to her that the bulk of this story was untellable. He could not convey—no one could—the tenderness, in fact the love, he had felt from this woman. More importantly he could not convey the feeling of having carried this memory across the distance of time, not just two years but the time of his life. She loved him for it, it was a good story, but one which another story almost immediately turned into something troublesome and ambiguous. It was as if he could not help himself, as if time swamped him with undifferentiated waves.

"There was a contest once at work, at the research institute, not an actual contest of course but something the research assistants cooked up on a long afternoon, something naughty and funny and weird."

He leaned toward his wife as he told the story. He did so in the annoying way some men have of implicating you in an intimacy and frivolity that isn't really there.

"Linda came in to ask me if I would judge a beauty contest and I could hear the others laughing, all the research assistants, in the hall-way outside my office."

"For Christ's sake how many of them were there?" his wife de-manded. She was annoyed with his raconteur quality. It cloyed at her.

"At least fifteen," he said. "It was a very large research institute. I was responsible for all the research assistants and the administrative assistants as well."

He knew something was wrong. "Is this too boring?" he asked.

"Go on," she said.

"It was a goof," he said, "one of those things that happens around an office on a slow afternoon. I said alright. Sure, I said. I'd judge their contest. Linda had a way of making you feel a fool if you didn't go along with something, and the others were laughing out of sight in the hallway."

She had handed him a stack of papers, photocopies really, blank side up. Each photocopy had a number written in pencil on the blank side. He was supposed to judge which image was the most beautiful and tell the number.

"It's like the golden apples," Linda had said.

He hadn't been able to place the allusion then. He knew it was mythological. He was somewhat wary but he agreed nonetheless and she scampered from the office, laughing. It was truly a scamper, something she wasn't given to normally. Then everyone ran away laughing, all the research assistants, as he turned over the pile of photocopies.

Each was an image of a woman's pubis, the wedge of hair and thigh and more or less belly depending on how the woman had arranged herself on the glass surface of the copy machine. Some images included the delicate fold of the woman's navel, some did not. The belly buttons were like rosebuds.

His wife found the story disconcerting and yet oddly fascinating as well. She had a vivid sense of what it must have been like for them to mount the machine and feel the heat of the light crossing over beneath the surface of the glass. They were girls in the city. Laughing. The whole scene vaguely gynecological as much as exhibitionistic. A vague sense of the sorority or convent as well. She could imagine this happening.

"The person they chose to get the results from me was the same woman who had helped me," he said. "The saint. She came to my office and calmly asked me for the number of the winner, collected the papers, and then left. No one told me whom I had chosen."

The photocopied images looked like nothing less than angels wings or the shadows of great moths. They had held themselves tightly upon the glass. There was very little anatomical detail although each pubic hair and each freckle or skin blemish was etched in detail. Even so the overall sense was one of holy beauty and simplicity.

SIXTH MEDITATION
Space (and Time)

I am as guilty as anyone of talking too much about it. A literary critic in Hamburg, Germany, once could not help sputtering as he accused me of valuing space too much over time. Space, space, space, he said (the repetition paradoxically turning the word spatial), stories are temporal. For him it was a sinful thing to think that a writer would lose track of mortality. In this last matter I agree with him, although I still am not certain that a devotion to spatial matters and an attention to mortality are incongruous.

"What time is it, I mean to say where am I," writes Cixous, "I mean to say where have I gone—I don't know anymore, in this instant when I call out to myself, where I'm passing or where I'm going."

For awhile there was a great deal of talk about surfing, although it has died down by now. Like most people my surfing experience is restricted to body surfing in uncertain swells along fairly safe shores. Even so the experience of being held by water, cupped in a wave and thrust, or ground down against a roil of sandy bottom, is like death itself, especially for a young man. Occasionally a strong wave punches the breath from you and then will not let you up and all of you aches to release into the force of it. This, of course, made it a bad metaphor for computer behaviors, although most likely this was a matter of perspective, the virtual surfer imagined in third rather than first person, seeing oneself in a mirror.

Sex is like that of course, subject to shifting perspectives and subjectivities. Yet it is more productive, more interesting, to consider the experience of therapeutic massage in a situation which is clearly established as nonsexual if never unerotic. There the self flows in and out of emotions which on other occasions we would clearly recognize as sexual, friendly, healing, familial, or happenstance. These emotions obviously have a temporality, they succeed and overlap each other throughout the experience; yet their experience is clearly spatial, both in one's own space and the space the masseuse occupies as she moves about the table, shifts a limb or head, hers or yours.

"Love comes like this: to the interior" says Cixous, and later, "the interior is so vast that an exterior is infinitely distanced outside time and space."

Once I swam out through low breakers and calmer, deeper water to a sandbar on which a hundred people frolicked or merely stood in the sun, miraculously in the middle (or what would serve as such) of the ocean. It was a long swim and I was tired when I reached the sandbar, a little wobbly on my feet, my biceps aching and dull, my eyes squinting in the salt glaze.

Before I could regain my breath, however, a lifeguard suddenly appeared in a long rowboat with a high, sharp prow. He was warning people through a coxie's megaphone. The tide is rising, the sandbar will disappear, you must go to shore now.

I looked at him with disbelief. My recollection is that our eyes met directly and he pronounced the warning again. All around me people were heading in. They began to swim in the force of the already rising tide, bobbing in the bright water. Their voices bounced off the surface of the water and multiplied in the waves, sounding like distant echoes, making me feel dizzy and sick to my stomach. I could hardly hold myself up against the heavy tide, hardly propel myself shoreward against the low weight of the swelling undertow. I thought I would drown. I watched the lifeguard rock in the swells in his row boat.

"A suave concavity spreads out," Cixous says, "the world is entirely hollow. We are equally hollow."

When I made it to shore I was exhausted and frightened and I laid there a long time next to my girlfriend on the blanket, unable to catch my breath or speak. My tongue was swollen and my ribs ached and I could hardly see through the hardened salt. She was annoyed with me for my silence and my inability to explain it. I thought of Lot's wife—an inversion I know—and space and time.

THREE LAST PIECES

Real Life

What she knew she thought she said.

It was a sort of motto for her. In some earlier time, say when her aunt Arlene was her age, she might have embroidered it upon a sampler.

"A sampler? Jesus!" Arlene laughed, "Have you any sense of history? At your age I'd put a dot of perfume behind each ear and another on each pillow and sink myself into a sailor. The war was over. There were silk stockings and a hundred brands of cigarettes."

Well one of those embroidered pillows then, the kind you see in movies.

"Whatever," Arlene said, she liked to affect the latest lingo.

"What she knew she thought she said," Elaine had typed it into her profile and her sigfile so it showed up any time anyone looked her up on-line and after her name any time she sent email.

"I suppose it means something," said Arlene. She was known as a kidder.

Elaine was embarrassed to admit she could no longer parse it exactly and so she didn't say anything. When she had typed it out the first time it seemed like a set of chinese boxes, one meaning inside another. Now she was confused.

She said what she thought she knew was one.

Arlene still smelled of cigarettes and perfume; Elaine tried to imagine sinking into a sailor.

"There was a film called 'The Sailor Who Fell from the Sea'," Arlene said. She was impatient, she wanted to see something.

"There," Elaine obliged her. The image loaded so slowly it made Elaine feel as if she were to blame. She wanted to apologize to her aunt who was seemingly mesmerized by the slowly focusing scene.

It was some girl's bedroom, far away in Oregon. Frilly bed. Stuffed animals with blue ribbons at the neck. Posters on the wall, standard issue. There was no one there.

"It is updated every five minutes," Elaine said, "Day and night. If you want, you can turn on streaming video and see it all happen in real time."

"Why would you?" Arlene asked. "Show me the girly shows. Bring on the live sex acts in cyberspace." She laughed wickedly.

Elaine hid the program and the screen returned to the orchids of her screensaver.

"I'm sorry, Arl," she said, "I'm not up to the joking just now. It's interesting is all. Glimpses into people's day-to-day lives. It's like you can see time passing."

"Or at least slow it down, kid," Arlene said. She had always called her kid, for twenty five-years now.

The girl in Oregon—her name was Lauren, Elaine knew— would be coming home soon. She would sling her things on her bed, a backpack with a leather bottom like a baboon, sometimes a portable CD player, a bottle of water, always Evian.

They could sponsor her, she could work a deal with them to sponsor the webcam, product placement.

"You mean the bottom of her pack is actually as red as a monkey's ass?" Arlene asked.

It gave Elaine a start, she hadn't been aware that she was talking aloud.

"Most days she tosses her bra on the bed as well. It's a virtual anthology of them. She fancies satin and pastels: pink, blue, green like Necco wafers."

"Do they still make Necco wafers?" Arlene said. "Does she think it's enticing?" she asked, "All very video queen?"

She could just be tired, Elaine thought.

"She could just be tired, some women hate wearing a bra. I do."

"I was fortunate enough to have lived through the sixties," Arlene said, "So have you met anyone yet?"

Arlene believed Elaine could meet someone new online. Arlene had read tales of chatroom encounters, email romances, Instant Message infidelities. Arlene was seventy-one years old and still getting laid, still sinking into sailors, albeit the kind who docked at the yacht club. Derek had been Elaine's first boyfriend, her lover, her husband, her betrayer, her ex, her slime-ball auld lang syne cop-a-cheap-piece mournful supposedly reformed old yeller red assed late Saturday night without a date asshole.

"Go girl," Arlene said.

She must have been thinking aloud she guessed.

Arlene messed with the mouse and Lauren's room appeared again.

Arlene and Elaine.

It was a difficult situation, their names so similar that in company one would reply to a query addressed to the other. They didn't go out much. Lauren was always going out, always sitting down to her girly-girl vanilla frosting vanity and painting her eyelids frosty blue. She would sit there in her pastel bra for the benefit of her webcam friends and prepare for a night on the town and on the bed with Keith, her skinny-assed (Elaine knew, she had seen them fuck on the webcam) boyfriend.

"Once Lauren kind of spanked him," Elaine said.

"Who?"

"Which one? Lauren? Or him?" Elaine asked.

"Which one what? This is like 'Who's on First?'" Arlene said.

"What?"

Communication is hopeless, Arlene did not say. She merely sighed as if to say so.

Tell me about her. Do you remember her getting ready for dates. Of course I do kid, my god she was so beautiful, so much the big Sis. I'd sit and watch her paint her lips red as your net friend's backpack. And she used powder, kid, wonderful stuff that smelled like lavender and went on like silk. And her boobs could fill up a slip, let me tell you, they were—how do you say?—ample. Not that she flaunted them, you understand, she was the essence of modest, a nifty kid, a real sweetie. My gosh, kid, when she smiled ice would melt in the ice box.

"I write to some old goat on AOL," Arlene said. "He's always asking me what I'm wearing. He has a big boat he says, a sixty-five footer or something, down in North Carolina or up in South Carolina or some goddamn thing. Says he hasn't had sex since his wife died but he's ready for me.

"You know what I remember best, kid? The nights we sat on the porch, the two of us, waiting for her date to show up, the crickets singing and the light posts spilling a cone of yellow light, smelling her next to me and hearing her singing to herself. And before long here he comes, you see him round the corner moving in and out of the cones of light like a relay. It was so good to see him, curly haired and handsome, his nickname was Curly and his hair shone in the street lamps. But it also made me sad, you know, knowing I would lose her. Aw shit, forgive me kid, sometimes I don't know what I'm saying."

Sometimes when Keith showed up he just wanted a hand-job. He would say that, "Come on babe, for old times' sake, just a hand-job." Sometimes it was easier to assent than to face up to him. He'd fall asleep and Elaine would wash off then move to the couch and wrap herself in a comforter like a taco.

What she knew she thought she said.

It hardly bothered her really, it was disgusting but she wasn't even fazed by it. He was so sad, really. The next week or so she would hear nothing from him. He'd be out with his cheerleaders and office girls, out with his gogo dancers and accountants, out with his policewoman and the woman who drove the donut truck at the construction site.

Sometimes Lauren wrote in her diary, one of those things with padded covers and a locking clasp. I swear she bought her furniture and her outfits from the prop room at Nickleodeon. She was straight sit-com. Still it was something to watch her as she wrote in her diary, her fist curled around the pen like a sea slug, biting her lips, her eyes dreamy.

What could she write there that we did not see? What secrets did she not share?

And what was it like then? I mean when she—

I cried my eyes out kid, I swear, I cried my eyes out. I held you in my lap and I cried my eyes out. I held you and told you I would never leave you, told you I would never let you be alone. I cried my eyes out, I cried my eyes out.

The official reason for their divorce was Keith wanted to have kids and Elaine didn't. You could ask him. If he was getting a little bit on the side that didn't really matter. That wasn't the point. He didn't want babies with *them*. She drove him to it in a way. A man wanted to have a family. In the old days you could get your marriage dissolved, without a divorce, if someone didn't want to have kids after they told you they did before you got married.

Arlene said she never got married because she was having too much fun. She had loved someone once, so deep she still mourned him, a fellow who owned an auto agency, that was the term she used, an auto agency. She meant a dealership. He wanted her to marry. He used to wear summer suits of seersucker and vests in winter. He took her to Paris and called it Paree, just like a sappy movie, although he was kidding when he did it. He died of a heart attack on the floor of the auto agency, the actual floor, the marble floor where they parked the Buicks behind the Art Deco windows in the days before the dealerships moved into pole barns along suburban strips. By the time the rescue squad got there he was blue, blue and dead like that. He left it all to Arlene, she was his one true love, and she never really had to work again, just like that.

"You weren't my age when the war ended. You couldn't have been more than twenty."

"Nearly eighteen, kid. Do the math. Eighteen was older then, you got five years credit for living through the war."

And she would have been twenty, Elaine's mother, Arlene's sister. She worked during the war at the ration office and never took any of the opportunities that came with that. She was "scrupulously honest," Arlene said, "almost virtuous."

Then she would laugh wickedly. "*Almost* kid, almost virtuous. It was a long war for a girl on the home front and boys came home hungry for cream."

Arlene was what would have been called a dirty old lady in that time. Now she looked ten years younger at least, her face and her breasts still smooth as cream.

"What's it like for you?" Elaine asked suddenly.

There was a light showing in Lauren's room on the screen and the sound of her flopping about off-camera. The two women watched as if something would happen.

"What?" Arlene asked absently.

"What what?" Elaine said.

"What it? What it do you want to know what it's like?"

Lauren smiled at her many net friends, her wide face grinning too close before the webcam like a pumpkin.

"You mean a jack o'lantern," Arl said.

She was talking aloud again.

"She seems like a hair brain," Arlene said.

"An air head," Elaine said.

"Whatever," Arlene said, "A hair head."

They laughed. Lauren was laughing too. It was a nice moment. What she knew she thought she said.

Their heads were close to each other, Elaine could smell the scent of her aunt, a perfume with the word *noir* in its title she guessed. Noir or nights.

"I am happy to live in a technological age," Arlene said too seriously.

Lauren wasn't doing anything to prompt such allegiance.

"What?"

"Vibrators," Arlene said. "I lived long enough to have a purse-sized little rocket. These old sailors can't quite dock it like the young men could. It's good to have a little friend on the bedside table."

Elaine was deeply embarrassed. She knew she was blushing crimson because she felt the flush.

"Don't tell me you don't use one," Arlene said, more shocked than incredulous, "Did you ever see the ad that says 'We put more women in orbit than NASA?'"

I don't read the papers Elaine thought.

Arlene laughed so hard she knew she must have said that aloud too.

Lauren was talking to someone on the cell phone. This was a new feature now that Lauren had added a members section to her site. Members could call her, or maybe it was have her call them, for an extra fee. It wasn't sex talk. They could just call when she was online and have the satisfaction of knowing she was talking to them while everyone else had to watch.

"I'll bet they jerk off all the same," Arlene said.

This was getting serious now. Either Elaine was communicating via ESP or she was talking aloud and unaware to some dangerous degree.

"That they do to the special pictures. There's a gallery of Lauren in the shower and Lauren wearing her teddie or G-string."

Arlene snorted. "G-string? G-string! How retro. I thought it was all thongs now. A panty line never worried me so much I wanted to run a thing up my butt crack."

Did you and mom talk like this.

There was no answer. Perhaps she hadn't said it, perhaps she hadn't thought it loud enough.

A very strange thing happened then. A man, Elaine had never seen him before, appeared in the window of the apartment across the courtyard on the same floor as her apartment. She had looked out the window a million times and never seen him before but it wasn't like he suddenly moved in or something. It was clear he was moving around a settled apartment. There was the usual mess men make but also a sense in the furniture and what-have-you that the place was settled.

Suddenly they were both watching him, she and Arlene. He wasn't handsome and he wasn't homely either, and he surely wasn't ugly. His sideburns were a little too long for his face and his hair was thinning at the crown enough that you could see the shine of his skull when he walked too close to the overhead fixture, a trio of bulbs in a half shell without a globe or a glass deflector.

Elaine guessed he was forty, Arlene said younger, thirty-five at the most. Lauren was still talking to the member on the phone but she was also unbuttoning her skirt which she let fall to the floor before folding it with her free hand, standing there before the webcam in her blouse and panty hose and looking like a sausage.

"Her legs, that is," Elaine explained, "Sticking out like that from under the tails of her blouse, they look like sausages."

"Still I'll bet the guy on the other end of the line is creaming his jeans," Arlene said.

The guy across the courtyard was eating a cheese sandwich. They had seen him make it, putting it together like a card trick: two white pieces of bread, a yellow square of cheese peeled from its plastic wrapper; stacking them like that, white bread, yellow square, a smear of very yellow mustard against that yellow, another white square. They watched him make the sandwich in one window and then watched as he disappeared and then reappeared again in another, turning on the light when he entered.

He, too, took his pants off. It was too much really, they both began to laugh, Lauren on the screen and the man in the window both standing on sausage legs, their shirt tails over their thighs.

The man in the window stripped off his shirt. His undershirt was very white. He held the cheese sandwich aloft while he stripped the shirt off.

She and her aunt were giggling aloud.

The man folded his trousers and laid them across his bed like a shadow puppet, the shirt he tossed in a corner where it looked like there were other shirts and laundry lounging against the wall.

"Isn't he going to drink anything with that ghastly cheese sandwich?" Arlene said. "I mean he'll choke to death. He might as well be eating his shirt."

Lauren had this disgusting habit of digging a finger into her navel, dressed or undressed, like she was burrowing for lint or trying to untie her umbilical knot. She was doing it now through the shirt while she talked on the cell phone to the lucky member.

Arlene howled with laughter. "The lucky member indeed!" she said and howled again, almost unable to catch her breath.

The man across the courtyard sat down to a table in his bedroom and started up his computer. Elaine could make out the Microsoft Windows logo from that distance though she couldn't really tell what came up next on the screen, probably email or a spreadsheet she suspected.

It was vaguely exciting and very sad to watch him, his back now to them, his face to the screen, a little white corner of the cheese sandwich on the table next to the monitor.

Lauren was crying. They had missed something. She reached toward the webcam and suddenly the picture went dead, just like that. Her mother has died Elaine thought.

Her own mother died when she was seven. Her aunt Arlene was the one who told her, who held her, who said she would never leave her and she hadn't yet.

In her mind Elaine believed she hadn't really cried for her mother, hadn't really cried at all she thought until years later when the space shuttle blew up in a split and forlorn stalk of white smoke when Christa McAuliffe died. That was 1986 and Elaine would have been eleven years old, six years after her own mother died. She had sealed the tears inside her like an amniotic sack and carried them for years. That year she cried. It was the end of January when the Challenger exploded and she cried all year.

The first American woman in space was Sally Ride and she was still alive. She too had ridden on the Challenger but years before it blew up.

Ride Sally Ride they said.

There was a song like that, it had to do with sex she thought.

Sally Ride was not the first woman in space. There was a Russian named Valentina something or other who went into space in 1963. She orbited the earth forty-eight times.

Sally Kirsten Ride was her full name. It was a nice name.

"I'm impressed, kid," Arlene said. "You know a lot about it, I'm impressed."

She said that but she was crying for some reason. Lauren was crying, Arlene was crying. Only Elaine and the man at the computer across the courtyard were not crying at this moment, at least as far as she knew then.

The Persistence
of the Ordinary

The reader may choose to mark it as either rhetorical wiliness or recklessness that this essay begins by saying it does not mean to make sense. And yet, in so saying, one means to offer neither an excuse nor outright encouragement for the reader to slip out now before it is too late. It has long been too late to make sense, although our current cultural circumstance, for which the internet stands as synecdoche, makes us belatedly ache with a longing for a lost coherence. We sometimes imagine our loss in terms of a golden age of literacy, for which reading or its imagined decline stand as silent witnesses of what never was.

It *is* late, perhaps too late, the revels over, and we are alone with what we have left after the stories are told and the musing is done.

Thus what is meant here is an act of witness as well, and so likewise one of reading. Here an extended meditation on willful incoherence is itself offered in the form of stories about different kinds of readings more or less intentionally formed without proper understandings. The readings involve different media, from book to film to video installation to the author's own supposed domain of hypertext.

Since already surely some readers may mark this kind of attitude and intention as postmodernist, it would be well to avow as much with a text from the postmodernist theorist, Michel de Certeau who suggests that "to read is to wander through an imposed system

. . . of verbal or iconic signs . . . a reservoir of forms to which the reader must give meaning" and in which "the reader takes neither the position of the author nor an author's position [but] invents in texts something different from what they 'intended'" and thus "detaches them from their (lost or accessory) origin" (169).

That a reader might differ with this claim in general one can expect and, as the reader might also suspect, this essay very nearly depends on it. Yet as warrant for the particular truth of de Certeau's claim I can and do offer my own detached origins and invented intentions.

At the end of the millennium in the fall of 1999 on a flight back from Milan after giving a talk regarding electronic literature, I was reading a paperback copy of Georgio Agamben's *The Coming Community* in the midst of waves of laughter from fellow passengers viewing the Hugh Grant and Julia Roberts comedy on the video screens. I gazed about the sea of serenely illuminated faces. Each was haloed by the dark rings of the earphones, whose foam plugs had the effect of keeping them from the experience I had of them as a composite audience. They were laughing to themselves in the same way I was reading to myself. To me, haloed as they were, they seemed like flights of angels.

Reading Agamben I was filled with delight and wonder alike and I had a dim sense in his figure of the quodlibet as threshhold; that is, "the experience of being—*within* an outside," of what seemed to me a hopeful model for my still only half-thought-through ideas for a study of virtuality and ordinary life and how the unseen becomes the unscene. That is to say that the mediated locale—where what is unseen is not the invisible, but rather, like silence, the setting of an unplayed scene—has nonetheless increasingly become where we locate our actual being, whether in a film or airplane or idea.

Caught up in this unthought I begin to wonder what to make of a mysterious and compelling paragraph in Agamben's chapter on Russell's paradox where he suggests that

According to a Platonic tautology, which we are still far
from understanding, the idea of a thing is the thing itself;
*the name, insofar as it names a thing, is nothing but the thing
insofar as it is named by the name.* (76.7)

Was there any relation other than foolish symmetry, I began to
wonder, between Agamben's notion and a recent and equally un-
thought-through claim of mine to my hypertext theory students that
"Just because something means what it says, does not mean that it
says what it means."

As best I could tell in the unmediated relative silence of the
cabin—where I sat unadorned by the foam earphones and only really
looking up at the screen, at least at first, when prompted by occasional
laughter or my own confusions as I read Agamben—the film (it was
Notting Hill, of course) seemed a mythic fable of mediation.

Without the soundtrack to guide me I began to develop a reading of the images which flickered over the horizon of the book. In
that reading this foolish film seemed not unlike Joyce's *Ulysses,* at
least to the extent that it suggests that the book and the image are
intimately (even erotically) linked through the everydayness of life
in which the goddess descends from her Olympian status as image
to consort with mere mortals within the span of a certain bookish
normalcy.

Yet in the film's unfolding we come to witness that normalcy is
increasingly exposed in the media, laid out in the pornographic sense
of the tabloids or their progeny, the voyeuristic webcams both pro-
fessional and amateur. Thus in a fairly iconic, even ideogrammatic,
sequence of scenes at one point in the film, a montage of tabloid pages
was followed by a literal mob scene depicting the predatory media
horde outside the hero's abode. The scene seemed to suggest—if a
bludgeon can be said to suggest—that media tantalizes us, displacing
gratification at exactly the point we come to experience it, leaving the

129

promised immediacy of original relations with the goddess thus un-consummated.

We are instead kept in a state of constant visual arousal under the sign of what Jay Bolter and Richard Grusin call "the double logic of re-mediation." The newest media, a la McLuhan, make the old their sub-ject, while the *soi disant* old media incorporate the surfaces of the new media, keeping us suspended between what Bolter and Grusin term "immediacy" and "hypermediacy," between presence and nextness.

Or so I thought seeing only the flicker of the images and adrift on others' laughter.

I began to watch more closely, although still not listening (in fact I have not since seen the film again either with or without its soundtrack). Instead I spent my trans-Atlantic interval away from Agamben attending to the images and trying to sketch, by way of tak-ing minutes of our airborn proceedings in my journal, the form of what I thought I saw there.

What follows next is thus a close reading of something I didn't actually watch all that closely. For instance a Vassar colleague, film scholar Sarah Kozloff, generously points out that in the summary of the film which follows I managed to miss an extended scene of what she characterizes as a tour de force of admittedly unexpected cinema magic. Therein our mortal Hugh rushes through a market lane in a shot "composed as if it were all one continuous take, one long track-ing shot . . . [with] a disguised cut about two thirds through" and which unfolds as a hypermediated time tunnel, its mise en scène a real time stroll through the ages of man where a pregnant woman at one end of the alley is seen midway with an infant and by the end with a toddler and so on for other characters.

In what I *did* see, and reported in the unauditory if not unau-dited redaction of my journal, the hero seemed to venture at end into the actual spectacle (a kind of Merchant Ivory romance, a film within the film, to which his red-headed and would-be beloved had fled like

Fowles's *French Lieutenant's Woman*). Yet from what I could see the hero's attempt to penetrate the spectacle of the movie within the movie and touch the screen goddess (even with words) could not withstand the spectacular intensity. So it is left to the goddess, as she often did in ancient myths, to descend again into the mundane space of the telling (seen in the film as the bookshop).

There, even if she cannot assure the filmic hero exactly, her descent does reassure the witnessing audience (i.e., we her chorus) that despite her seeming distance from us she is nonetheless one of us; that she hears our prayers of longing, adorning our ordinary lives with glamour and laughter; that she recognizes and seeks us in our insulated solitude, a solitude she mirrors for us in celluloid remove and outside language, and at which distance deigns to dwell among us at least for the span of our airborne, screenward reveries.

In the film's penultimate scene a car chase ensues as the hero in company of his ordinary companions pursues the fleeing goddess, Chaucer's pilgrims alle headed now toward some new, defrocked Canterbury. Even a bloke on crutches, one of the hero's pals, throws them down and takes up the chase as if to affirm that toad's wild ride miraculously includes us all.

In fact the hero's mates seem to surround him as something of a newly reconstituted family, a composite clan which offers him and us a renewed promise of crossing class boundaries and thresholds alike as if bound up in a single silver cylinder hurtling heavenward. All together now we are increasingly caught up in the mad chase and hell-bent toward crashing the Olympian party and along with the hero and his clan pressing the case for the hero's worthiness, an imbrication signaled by the accelerated waves of raucous laughter through the cabin, that laughter bobbing along both at the tempo of the car chase inside the film and above the steady droning of the jet engines in the high cold air outside it.

Just at the moment the movie's traditional comic banquet

threatens to become a drive-through, our pilgrim band tumbles out into the mediating presence of the transubstantial press conference. The scene is a sort of epithalamium, a prenuptial ode in the form of a fairy tale in which, given one question, the classic portion of the suitor in fables, our hero (by now—as indeed he always ever was from the first—himself only a screen image of himself, Hugh Grant playing a man very like Hugh Grant but caught up in tabloid mysteries in the way of, well, Hugh Grant) appears like a man in a hall of mirrors, framed in the film frame within the press conference television monitor, on the cabin screen, and in the dead center of the aluminum tube hurtling above the clouds over Newfoundland.

So skewered, confined, and defined, and thus, for a brief moment at least, not unlike us, he gives up his ordinary and bookish life in order to redeem us from the eternally distancing image.

Yet we can none of us nor him escape what frames us (this is an old lesson of comedy, from the Greeks to Shakespeare). Agamben picks up on the etymology of framing,

> the notion of the "outside" is expressed in many European languages by a word that means "at the door" (*fores* in Latin is the door of the house, *thyrathen* in Greek literally means "at the threshold"). The outside is not another space that resides beyond a determined space, but rather it is the passage, the exteriority that gives it access—in a word, it is its face, its eidos.
>
> The threshold is not, in this sense, another thing with respect to the limit; it is, so to speak, the experience of the limit itself, the experience of being-within an outside. This ek-stasis is the gift that singularity gathers from the empty hand of humanity. (67)

Ek-stasis, stepping out, high stepping, is likewise the end of comedy and how comedy ends from the Greeks to Shakespeare to

Wonder Boys. And so it was likewise with the mute and high-flying comedy I recount here whose sunny countenances rose above and set below the horizon of the book before me.

At the conventional comic banquet which seemingly ends the film, all the couples kiss conventionally according to class and station—whether regal, sylvan, mechanical, first, business, or economy class—the hero and heroine lingering briefly in slow dissolve on paired screens like a double cameo, signets of an empire of signs.

Yet like most media productions in our time this film is not content with a single ending but rather dissolves itself in a series of endings. If as Agamben suggests passage gives access to its own face, this passage is at least two-faced. What seemed a soft-focused double cameo screen of goddess and her hero turns out to be merely the rear window of her limousine. And as the duo are driven out of this comedic utopia the wave of the bridal starlet gives way to a soft-focused pastoral envoi, a return to the garden.

We flash forward in soft focus to an Edenic gambol of boy and girl as Blake's innocents whose fluid innocence leads the camera to the slow reveal of the hero and dismounted Olympian heroine, themselves newly innocent within the garden, Birkenstocked and besocked, hands clasped in reverie and the book respectively. Or so I saw, without hearing.

A reader may now wonder, beyond some winking irony, what this impossible perhaps even irresponsible reading is meant to suggest. Is it perhaps a pale parody of the kinds of close reading and theoretical framing which occupy us in what's left of the profession of literature?

I have little to offer by way of either answer or justification other than my particular interest in what people ordinarily think as they encounter narratives, an interest that inevitably extends itself to how or what or whether they ordinarily think of images and texts in relation to themselves and the other media which increasingly constitute, or crowd, their lives.

I am a practitioner and a professor alike of a literature that has been variously described as "Myst or Warcraft II as re-imagined by Robbe Grillet" and which "no one really wants to read . . . , not even out of idle curiosity," one which displaces plot and character in favor of "the listless task of having to choose among alternatives," and which is "entirely inhospitable to the more subjective materials that have always been the stuff of art. That is to say, . . . antithetical to inwardness." These are characterizations which arise from Michiko Kakutani, Laura Miller, and Sven Birkerts respectively (which in a snit of tit-for-tat you will not find listed among works cited) and whose own reveries, one might suggest respectfully, are the critical equivalent of seeing a film without a soundtrack, that is responding to a literary form without ever quite reading its, admittedly unstable and multiple, texts.

The question increasingly before us is quite literally what we make of such viewings or readings. People have moments as ornate as mine on airplanes under the influence of contemporary neoplatonists; others rise and fall in laughter and reverie in the empty hours between three-quarter-scale Disneyfied and dyspeptic minimeals in economy seats. Both may be equally foolish, yet both describe the persistence of what we mean by our humanity in an age of ever newer media.

Our task is how to link experiences ceaselessly battered by the progression and promise of what flickers in the dark.

Our locale is ever more Agamben's threshold, for in our mediated lives books, movies, and hypertexts alike veer ever toward "the experience of the limit itself, the experience of being-within an outside."

A few weeks after my airborne quodlibet I had another threshold experience when I finally caught up to the Bill Viola show at the Art Institute in Chicago.

I say finally because I had missed its original run in New York the previous year at the Whitney during one of those winters and springs when the city became unmoored and, as it does periodically, floated out to sea, suddenly making for a time unreachable what was normally only an hour's train ride away. Luckily however, as already noted here, most media productions in our time—including art exhibitions—are not content with a single ending but rather dissolve into a series of them.

This of course makes every ending somewhere a beginning elsewhere. And so although by the time the floating city returned to its dock the Viola show had detached itself and drifted farther east, riding the jetstream over Europe like a soap bubble or a dirigible, touching down in Amsterdam and Frankfurt, until drifting again over the western ocean to San Francisco, and eventually across the inland ocean to Chicago, there it caught up with me or I with it.

Throughout that show, as with most of Viola's work, the expectation of event plays off our own weariness with event. One first enters a high dark hall into the presence of a work called "The Crossing" commemorated in the show's teeshirt and catalogue and whose iconic messenger roars with fire and water so loud it can be heard galleries away. This portal work was a banner of light splayed out in the cathedral darkness like El Greco's great altarpiece "The Assumption of the Virgin," which hung elsewhere in the art institute.

To see Viola's work is to find oneself similarly and repeatedly at a crossing. Again and again, from the exhibit's first entry hall through each successive installation space we long to get on, to turn the channel, change the scene, a longing exacerbated by the seductive sounds of installations in neighboring galleries, many of which are connected by narrow, dark corridors like birth canals or hypertextual links.

Each time, just at the point we convince ourselves to tarry, just when we realize that this time is the time of its own experience, that nothing will happen except what we are now seeing, suddenly something terrifying, splendid, transporting, surprising, or transcendent truly happens and we have to move on.

"As we continue our dance with technology, some of us more willingly than others," Bill Viola once wrote, "the importance of turning back towards our selves, the prime mover of this technology, grows greater than the importance of any LSI circuit."

Against the implicit diminishment of the nextness that assails us we need to turn back upon the unique circuit of momentary recognition which rescues meaning itself from any fixation. What is fixed no longer, if ever, lasts. Each thing now flows.

Each glistering thing, both new and old, must be plucked from the scoured gravel we dip from the flow. From and in our technologies we need to fashion a mode of conservation which locates meaning itself likewise in the flow of remembered time.

We must insist that we find meaning in the quotidian where Shakespeare did likewise—and Emily Dickinson and Gertrude Stein and Kathy Acker and Joan Retallack and Shelly Jackson after him. We must insist that a living culture is measured in its ability to mark such passage with forms as dynamic as the art that evokes and sustains it.

The call for attention to the ordinary is not a new thing in our time, and has especially occupied French thinkers from Henri Lefebvre's *Critique of Everyday Life* (1947) to his disciple de Certeau's *Practice of Everyday Life* (1980). Yet it is a consideration that takes on a renewed urgency and importance in the face of new media, which by their nature dislodge the quotidian and diurnal, the day to day and the moment to moment.

Lefebvre situates the everyday "at the intersection of two modes of repetition: the cyclical which dominates in nature and the linear, which dominates in processes known as 'rational'." (36) In a darkly prophetic utterance he notes that "in modern life, the repetitive gestures tend to mask and crush the cycles."

In theorizing and creating early hypertexts many of us saw a form suited to reifying and retrieving the cyclic richness of ordinary life. To be sure the webbed network seems to promise a version of de Certeau's "network of antidiscipline" as a way out of the tyrannies of repetitive gestures. What none of us—modernist, Marxist, postmodernist, feminist, hypertextualiste—could quite anticipate was the current regime of repeated and insistent novelties. Devised to shrink the cyclic into the momentary, they deny and depose our sense of mortality, which is nothing less than our awareness of how the cyclic gives meaning to repetition.

The net dislodges the quotidian and diurnal by occupying it in every sense of that word, filling space and time alike. The posthypertextual commercialized webbed network advances an ideology of nextness in an explicit embrace of the proverb, which the protohypertextualist and that most cyclic of writers Gertrude Stein rejects: "Familiarity does not breed contempt," Stein writes, "anything one does everyday is important and imposing and anywhere one lives is interesting and beautiful. And that is how it should be"(18).

The fundamental familiarity of literature is its sense of life lived in common and commonplace. To profess the persistence of the ordinary in the face of constant nextness, we must set aside any fascination with the merely lasting in favor of conserving the lasting power of reflected and reflective moments. The hypermediated surface, the slow, tropic flow and swirl across the face of the soap bubble, is where we withstand the concurrent and concussive blast of immediacy, the onrushing nextness of unmoored life.

The hypertextual link, as any number of contemporary cybertext theorists suggest, severs as much as it links. It is itself a story, a moral tale and a meditation, a hesitation, a perturbation in the morphogenetic sense, a form-making shift in form, a curl not unlike a whirl of commas, which forms the flow of both what it links and what it sets off from and in. Its type and figure is the Bernoulli effect, the still largely mysterious phenomenon that follows upon the curl and flow of increased stream velocity in a fluid like air and that allows a wing to lift and keep us and our books and screens, our earphones and dirigibles, aloft over far seas in each other's solitary, momentary company.

The Future of
Fiction and Other
Large Phrases

The future of fiction is in its past and ours. Fiction is residue. Fiction in whatever form calls us to find purpose in surface, to discover and recover a lasting world in our shifting sense of ordinary life. Fiction is a summoning to what I have elsewhere called ordinary mindfulness, the web of caring with which we surround our day-to-day lives and those we love. Surface attentions grace human life and grant us the peculiar joys that actualize even increasingly virtual lives. As residue, fiction increasingly seems the sole mode of adequate representation, which is to say truth-telling, for a world in which the nature of change itself changes constantly. In the midst of an incessant procession of nextness the shadowy presence of our actual past disappears in the flare of the bright succession of creatures who shape our formless futures.

"The exterior of the city is no longer a collective theater where 'it' happens" Rem Koolhaas writes in *S, M, L, XL,* "there's no collective 'it' left. The street has become residue, organizational device, mere segment of the continuous metropolitan plane where the remnants of the past face the equipments of the new in an uneasy standoff" (514).

Fiction, which is increasingly seen—especially in cyber-schemes of so-called story worlds—as an architecture, is an architecture of residues, a pile of cinders in Derrida's sense of them: "there not here, as a story to be told: cinders, this old gray word, this dusty theme of humanity, the immemorial image had decomposed from within, a metaphor or metonymy of itself" (31).

"For the writer . . . it is less and less necessary . . . to invent the fictional content," Koolhas quotes J. G. Ballard (from the introduction to *Crash*) in the interlinear, marginal, and hypertextual dictionary within *S, M, L, XL* , itself a residue of a hundred texts, "the fiction is already there. The writer's task is to invent the reality."

We invent by interposing ourselves in every sense. Increasingly the only past we have is the residual image which we make up for and of ourselves as remnants. This making up ourselves, for instance, forms the clarifying shadow that introduces and insists on the meanings of our presence in the relational architecture work "Re:Positioning Fear" of multimedia artist Rafael Lozano-Hemmer, a collaboration with Will Bauer. Not surprisingly among the eight thematic sessions of this work was one devoted to "Fear at the End of Architecture."

The texts of these sessions, the already past (and passed, i.e., once "real time" IRC or internet relay chat) stories of the participants hopes and fears, were projected on the courtyard facade of one of Europe's largest military arsenals, the 350-year-old Landeszeughaus in Graz, Austria. "Even though the IRC sessions could have been projected on the arsenal by covering most of the facade," the project website explains, "an interface was designed to prevent all of the text [from being] visible at one time." This "tele absence" interface consisted of "an 'active' shadow that reveals the text on the building . . . [whose] final effect was a 'dynamic stencil' whereby the shadow of the participant . . . 'revealed' the IRC texts . . . within the building . . . as though the shadow was a cutout or an x-ray of the building."

The past matters, the past is matter, residue: household dust is sloughed skin, a present sign of our coming absence, the chaff of mortality. Dust, as Joseph A. Amato reminds us in a marvelous pun "not so long ago . . . constituted the finest thing the human eye could see." Even a small child knows that upon closer examination the blackest soil reflects us in a million anaerobic mirrors. The smeared light of computer screens and our skin have much less than six degrees of separation. Yeats's place "where all ladders start, in the foul rag-and-bone shop of the heart" sparks the alchemical wellspring for the brilliant hydrolytic chain whose alphabetic friction, from ATP to ADP to AMP, transforms the light of the world.

Light is made of words, and words are light's lone embodiment, the fiery comet's tail of Indra's skirt.

Why then is networked literature so surfaceless, so dull, I wonder?

In a grim, glib but apt *New York Times* review of the first internet art in the Whitney 2000 biennial, Michael Kimmelman says, "most of the sites I visited didn't amount to much more than scrolling texts, fuzzy pictures, remote video cameras and interactive gimmicks . . . slow to download and visually inert. One of these days the Internet will come up with something good."

Commentators are used to saying that such and such networked form, from e-art to e-commerce, has not yet fully emerged. Indeed electronic fiction has not yet emerged and yet will *never* emerge in a finished state in which we will suddenly recognize that *this* is where fiction has gone, this is what it has become. Shifting fictional forms undergo no holometabolous metamorphosis. They leave residue behind and more often than not monstrous creatures, half-transformations, flying carcasses, decaying pupae crawling through their own chaff on caterpillar feet.

It is this eye for the monstrous that I like best about Finnish novelist and media theorist, Markku Eskelinen. In his world, knights errant mount flying pigs with the radiant wings of moths. When he reimagines the MOO (online virtual story worlds) "from the perspective of Augusto Boal's Invisible Theater" where "participants do not and cannot know the boundaries separating the realms of fictive and real-life communications, or those between persons, actors and roles . . . [and where they] participate but they do not know for certain in what," he'll take the monsters at face value. He isn't worried about fully formed emergences or the circus animals' desertions.

Not satisfied by his own mentor Espen Aarseth's Nordic cybertextual forms and austere, arithmetic reformations, Eskelinen proposes "postnarrative" story forms of viral simplicity in which "the attitudes and speech acts of our real world are given their chances to inflict the fictive world" and where A-life "emergent traits" and "glider narratives" lead to "ecologically delicate islands getting easily off balance." On these asymmetrical Galápagos, "alzheimerian filtering" and "tel quelian . . . search engines" generate "kinetic textual dance" in the way that fruitflies' frantic dances spell out genomic poetry.

Right now we need a certain sanguine, consanguineous shaking up, *Un Coup de dés,* which might result in some monstrous turning. Web-lit in its fruitless quest for emergence has seemingly left off its mutant longings and however briefly fallen into settled forms.

I count no more than three at present, their several variations no more difference than the kinds of barbed wire in a cyber cowboy's collection up for grabs on e-Bay.

Too much of web-lit, including my own thusfar, consists of extended captioning of fey graphics, Kimmelman's mix of scrolling texts and fuzzy pictures, faux aleatory forms gussied up as Tarot masquerade. Click a pic and see a poem, click a couplet and see a coupling.

Secondly, speaking of couplets and couplings, there are solemn procedurals, jejune chat room assignations and dreary one-night teledildonic virtual stands, whose austere and flowery language games alike are carried on with ghastly seriousness. MOO-whoing, webcam keyholes, and bare-assed pajama parties pretend to be broadband jamborees but turn out to be only streaming idiosyncrasies sans idiots savants.

Finally and perhaps most promisingly, there's quite literally all sorts of poetry in motion, Kimmelman's interactive gimmicks in the form of oulipian lava lamps. Maybe something good will come of this babel of towering good will, postsyntactical repositioning, and Eduardo Kac-ian kinetics. Here at least monsters are possible, strange birds, Uirapuru stew and other singularities.

Michael Eisner killed Mr. Toad's Wild Ride, perhaps the best immersive and interactive fiction of the twentieth century two years before that century's now agreed-upon end on September 7, 1998, according to the elegiac website which collects "more than 1100 submissions from Toad fans all over the world."

It is too easy, a typical postmodernist trick, to move from this to a punning mention of Artaud's Last Ride. Done in by the director of French Radio, Vladimir Porche, Artaud died of censorship on March 4, 1948, a month after the retraction of his radio play "To Have Done with the Judgment of God" ("*Pour En Finir Avec le Judgement de Dieu*"). By all accounts its major focus is quite literally shit, that most basic of residues, of which he wrote once

Where there is a stink of shit
there is a smell of being.

According to Alan Wiess, the play was meant to end with the autopsy of God who appears as a dissected organ taken from the defective corpse of mankind. To a soundtrack of shrieks, screams, and grunts, Artaud had devised a glossolalia of nonsense sounds to mark the disconnection of meaning from language.

Why do I tell you all this? It is residue, a history of someone, a story. Your or someone's past. "When we speak the word 'life,' it must be understood we are not referring to life as we know it from its surface of fact, but to that fragile, fluctuating center which forms never reach."

Artaud wrote this in *The Theater and Its Double.* And in a poem called *Post Scriptum,* not just a PS, but after writing, he wrote this:

> Who am I?
> Where do I come from?
> I am Antonin Artaud
> and if I say it
> as I know how to say it
> immediately
> you will see my present body
> fly into pieces
> and under ten thousand
> notorious aspects
> a new body
> will be assembled
> in which you will never again
> be able
> to forget me.

It is there, in the post scriptum, in the writing after writing, that I choose to see the future of fiction, or at least as much as it is given me to see, in a new body beyond forgetting.

AFTERWORD

Hèléne Cixous

Yes, these narrative crystallizations or *precipitations* are reminiscent of James Joyce's epiphanies. The heart of each story is a coming-to-awareness. But with Michael Joyce, these epiphanies have an interrogative rather than a conclusive form. They tell of the emergence of moments in which the characters' lives are destabilized. These are "imponderables," instants at the edge, on the verge, near the end: of love. of life. Let us call them *tales of almost.* They capture the glancing contact of differences brushing against each other. Each one is a conjuring of those limit-states of consciousness, within the liminal location where thinking begins. As if writing were recording the sparks that fly as different states of consciousness strike against each other. And the reading in turn is set ablaze.

Flights of consciousness, "flights of fancy": at once flying and fleeting. Or rather what Michael Joyce calls "virtual realities": these must be understood as a new category of states between virtuality and reality, virtureality, the virtual *within* reality. We find ourselves at a point of articulation where the virtual and the real intersect. Remembering things that have not happened: these are made-up memories but not déjà-vu. What is at issue is the elasticity of time, as if it were possible to *space time,* that is, to space time out, to stretch time or to cut time with space. . . . The frame within which these stories take place is always that of everyday life such that the reader not expecting any surprises is caught unawares. For example, on one occasion a woman asks her husband, "Have you ever known a saint?"

(and here Michael Joyce plays on the indeterminate gender of the word "saint" in English). The answer comes on the same page, a few lines further on, two years later. The question has traveled round the clock. Round two worlds.

These philosophemes: the viscosity of time, the introduction of the fantastic into daily life, the uncertain limit between life and death are staged in a misleading reality. The banality of the localization is a simulacrum. It is at once deconstructed by a thought event, a mental sensation, not by a worldly event.

Michael Joyce is a subliminal explorer—he sets off to explore mental regions that are generally neglected, as if they were forests or deserted islands. Yet his stories are idiomatically dated to our time. From a temporal point of view he is an idiomatic witness to a particular American era. He consigns the phenomena of memory as a botanist who spurns sensationalism while gathering a herbarium of his times.

These absolutely original tales are at once idiomatic and unclassifiable. Slow, listless, yet incredibly speedy. Ripples allude to the deepest depths. This is the secret of great poetical writing.

Michael Joyce's humor moves from the everyday to the dead serious: the imminence of death.

What confers a distant music to all of these texts, which seem to be quite heterogeneous, is *love*. At times this love looks worn out like old clothes, but it is nonetheless the thread that weaves its humid way through the whole collection. Tears collect silently—smiling.

–Translated from the French by Eric Prenowitz

WORKS CITED

Agamben, Giorgio. *The Coming Community* (Theory Out of Bounds, vol. 1). Trans. Michael Hardt. Minneapolis: University of Minnesota Press, 1993.

Amato, Joseph A. *Dust: A History of the Small and the Invisible.* Foreword by Jeffrey Burton Russell and illustrations by Abigail Rorer. Berkeley: University of California Press, 2000.

Artaud, Antonin. *Four Texts.* Trans. Clayton Eshleman and Norman Glass. Los Angeles: Sun and Moon, 1982.

———. The Pursuit of Fecality, in *To Have Done with the Judgment of God* (1947; reprinted in *Selected Writings,* ed. Susan Sontag, 1976, p. 37).

———. Preface: The Theater and Culture in *The Theater and Its Double.* Trans. Mary Caroline Richards. New York: Grove Press. (1938; tr. 1958)

Beckett, Samuel. *The Complete Short Prose: 1929–1989.* Edited with an introduction and notes by S. E. Gontarski. New York: Grove Press, 1995.

Ballard, J. G. "The Thousand Wounds and Flowers." Review of J. T. Frazer, *The Voices of Time,* in *New Worlds,* 191 (1969).

Bolter, Jay David, and Richard Grusin. *Remediation: Understanding New Media.* Cambridge: MIT, 1999.

Borges, Jorge Luis. "A New Refutation of Time" in *Labyrinths,* New York: New Directions. 1964.

Calvino, Italo. *Six Memos for the Next Millennium.* Cambridge: Harvard University Press, 1988.

Cixous, Hélène. What Is It O'clock? in *Stigmata: Escaping Texts.* New York: Routledge, 1998.

de Certeau, Michel. *The Practice of Everyday Life.* Trans. Steven Randall. Berkeley: University of California Press, 1983.

Derrida, Jacques. *Cinders.* Trans. Ned Lukacher. Lincoln and London: University of Nebraska Press, 1987.

Eskelinen, Markku. 1999. Cybertext Narratology. Paper given at the Digital Arts Conference, Atlanta. Available from World WideWeb: (*http://www.lcc.gatech.edu/events/dac99/*)

Genette, Gerard. *Narrative Discourse.* Ithaca: Cornell University Press, 1983.

Hemmer, Rafael Lozano, and Will Bauer. 1997. Re:Positioning Fear, Film+Architektur Biennale. Graz, Austria. Available from World WideWeb: (*http://xarch.tu-graz.ac.at/home/rafael/fear/*)

Kimmelman, Michael. 2000. A New Team at the Whitney Makes Its Biennial Pitch, 3/24/2000. Available from World WideWeb: (*http://www.nytimes.com/yr/mo/day/news/arts/whitney-biennial.html*)

Koolhaas, Rem, OMA and Bruce Mau. *S, M, L, XL.* New York: Monacelli Press. 1998.

Lefebvre, Henri. *Critique de la vie quotidienne,* Paris: Grasset, 1947. Published in English as *Critique of Everyday Life.* Trans. John Moore. New York: Verso, 1991.

Panahi, Jafar. *Badkonake Sefid* (*The White Balloon*). October Films 1995.

Save Mr. Toad's Wild Ride! (*http://www.cs.miami.edu/~jam/toad/proj.html*)

Stein, Gertrude. *Paris France* . (originally published in 1939). New York: Norton, 1996.

Viola, Bill. *Bill Viola,* exhibition catalogue. Curated by David A. Ross and Peter Sellars. Whitney Museum of American Art, New York and Paris: Flammarion, 1997.

———. "Will There be Condominiums in Data Space?" In *Reasons for Knocking at an Empty House.* Cambridge: MIT, 1995.

Alan Wiess. "Radio, Death and the Devil." In *The Wireless Imagination: Sound Radio and the Avant Garde.* Ed. D. Kahn and G. Whitehead. Cambridge: MIT, 1992.